Antonia Purk

Jamaica Kincaid's Writings of History

American Frictions

Editors
Carsten Junker
Julia Roth
Darieck Scott

Volume 8

Antonia Purk

Jamaica Kincaid's Writings of History

A Poetics of Impermanence

DE GRUYTER

Dissertation accepted at the Faculty of Philosophy of the University of Erfurt in 2021.
Supported by Open Access funds of the University of Erfurt.

ISBN 978-3-11-101861-4
e-ISBN (PDF) 978-3-11-102750-0
e-ISBN (EPUB) 978-3-11-102752-4
ISSN 2698-5349
DOI https://doi.org/10.1515/9783111027500

Library of Congress Control Number: 2023938299

Bibliographic information published by the Deutsche Nationalbibliothek
The Deutsche Nationalbibliothek lists this publication in the Deutsche Nationalbibliografie; detailed bibliographic data are available on the Internet at http://dnb.dnb.de.

Printing and binding: CPI books GmbH, Leck

www.degruyter.com

Acknowledgements

This book greatly benefitted from the support I have received throughout the years inside and outside academia. My gratitude goes out to my dissertation advisor, Ilka Saal, for years of productive discussion and an exchange of ideas that inspired and guided me while working on this project. You taught me ask better questions and to never stop asking them. I thank you for your continuous curiosity and interest in my work, for your generous and critical feedback, and for having allowed me to learn from you. I owe many thanks to my dear colleague Verena Laschinger for her encouragement, inspiration, and friendship. You both are true examples of intellectual integrity and human kindness that I can only hope to live up to. I thank Ralph Poole for his enthusiastic support as my external reader and for his interest in my project.

I thank the University of Erfurt for supporting my research with the Christoph-Martin-Wieland scholarship. A big thank you goes to my friends and colleagues of the graduate group "Texts.Signs.Media" at Erfurt for our time of thinking, reading, and writing together. Our discussions at the intersections of different philologies have enriched my work, and our informal online writing group during the pandemic helped me finish my dissertation during times of isolation. My thanks go especially to Michael Friedrich, Felix Haenlein, Kerstin Howaldt, Johanna Käsmann, Tobias Schmidt, Elena Stirtz, and Alexander Waszynski. Kerstin, I could not have wished for a more delightful colleague to have shared an office and the experience of working on a doctoral thesis with – you always made any unfun things fun. I am grateful to the Harriet Tubman Institute for Research on Africa and Its Diasporas for hosting my research stay during an early stage of my project and granting me insightful historical perspectives on my study of poetics.

I am grateful the people who helped me through the final steps of turning my dissertation into a book. I wish to thank the American Frictions series editors, Carsten Junker, Julia Roth, and Darieck Scott for their interest in my work on Kincaid's poetics and for granting me the opportunity to publish my book in the series. At De Gruyter, I was fortunate to work with Julie Miess, who guided me through the stages of the publishing process. I thank Michael Thomas Taylor for his detailed and attentive reading of my manuscript during the copy-editing process.

My last words of gratitude go to my family. To my parents, Christl Maier-Purk and Klemens Purk, who instilled a love for texts and travel in me. From you I learned about the joys of seeing, reading, and tasting something new every day. My heartfelt thanks go to my partner, Michael. As my favorite interlocutor, he listened to every early thought and later read them on the pages of this book. Micha, your love, tireless support, and reassuring encouragement are everything to me.

 https://doi.org/10.1515/9783111027500-001

Contents

Introduction

"[F]or me history was not a large stage filled with commemoration, bands, cheers, ribbons, medals, the sound of fine glass clinking and raised high in the air; in other words, the sounds of victory. For me history was not only the past: it was the past and it was also the present. I did not mind my defeat, I only minded that it had to last so long," muses Xuela, the narrator and protagonist of Jamaica Kincaid's 1996 novel *The Autobiography of My Mother* (138–139). Xuela Claudette Richardson is the descendant of a Carib mother and an Afro-Scottish father; her narrative is set in Dominica in the early twentieth century. In sketching out the events of her life, the text presents Xuela as grappling with the history of colonialism in the Caribbean as it affects her in her everyday life.

Xuela's estimation of history immediately calls attention to the power imbalances of history. On the one hand, this pertains to the historical subjugation of one group by another. On the other hand, "history" as a particular way of perceiving the past is firmly tied here to the side of the victors. Looking from the outside in, Xuela determines that "history" is being *done* in the present. In this, it is not solely to be understood as a reference to the past, but also to contemporary engagements with past events. History in this case is located in its performance on a stage of commemoration, in material products, such as ribbons or medals, and in the enactments that produce a sense of celebration, including the playing of music or the clinking of glasses. As such, history is constituted in the present.

The nexus of entangled temporalities, (re)enactments, and mediality that Xuela here identifies as the defining characteristic of history are at the center of this study of Kincaid's writings. Asking what constitutes history and how literature intervenes in historical consciousness, I examine the media text, image, and the human body and their poetic and performative engagements with history in Kincaid's fictional works.

Xuela differentiates between "history" and "past" on a temporal scale. History is not equated with past, but it continues into the present, muddling temporal levels through a pervasive simultaneity. Her own history that affects Xuela's present differs from the history of the victors because of this temporal entanglement, which signifies an incompleteness when the effects of "victory" persist. She thus criticizes less what may have happened in the past than the continuation of defeat through the victors' version of history as it is perpetually mediated in the present. Xuela's critique points to a plurality of history – or rather histories – depending on the perspective.

Based on my reading of the above passage (among many others from Kincaid's works), I understand the "past" as something that has happened in an earlier time

 https://doi.org/10.1515/9783111027500-002

and to which we only have access through a range of media. This mediation can come as the commemoration made by the clinking of glasses or through written texts, or it can be sexual practices performed with human bodies. In each case, it is this medial fashioning that then turns the past into "history." This is what makes the past available in the present, and which might thereby also repeat and perpetuate past power structures in the present. As I use it, the term "history" thus refers both to a consciousness of the past and to the various medial forms in which it is articulated, and which produce this consciousness. History is a representation of the past, always already filtered by the ideals and standards of those who produce such representations. Xuela sees the victors, i. e., the colonizers, as those who determine representations of the past.

Speaking for a collective that is faced with defeat, Xuela points to the exclusion of these voices from history.[1] However, in laying bare the medial entanglements and the possible plurality of histories, and by inserting her own voice into this negotiation, Xuela questions the colonial victory over discourses of history. Here, *she* is the one speaking about her own and her ancestors' past, thus turning it into a form of history. And this is where the productive potential of Kincaid's writings of history lies: such a postmodern understanding of history and historiography exposes the intentional production of historical discourse., i. e., its "made-ness."[2] As such, it is fair game for anyone's interventions – including those of Kincaid's protagonists and narrators, such as Xuela, who are thus enabled to put forth their own versions of history, with their own perspectives and their own voices.

Towards the end of *The Autobiography of My Mother*, Xuela determines that "[t]he past is a fixed point, the future is open-ended; for me the future must remain capable of casting a light on the past such that in my defeat lies the seed of my great victory, in my defeat lies the beginning of my great revenge" (215–216). While the past may appear fixed to Xuela, I argue that in Kincaid's texts it is not. By looking at the past from a future that is yet to come, perspectives change, as Xuela suggests. A view that looks retrospectively from the future can see the possibility of victory and revenge in defeat. In this way, a back-and-forth motion in perspective destabilizes the fixity of the past – and even more so that of history – and decenters colonialism's victory over discourse.

My analyses of medial engagements with history seek to highlight the poetics of impermanence that produce such change in Kincaid's works. Through textual

1 For instance, about her lover Roland, an Afro-Caribbean stevedore in Dominica, she says, "[h]e did not have a history; he was a small event in somebody else's history" (*The Autobiography of My Mother* 167).

2 In my use of the term "historiography," I draw on Hayden White's thoughts on historical representation in *Metahistory* and Frank Ankersmit's *Historical Representation*.

strategies of ambivalence, referential redirection, repetition with a difference, and cyclical motions of narrative, the texts enact aspects of Caribbean colonial history in a way that creates multiplicities of perspectives, voices, and histories. In short, through poetic form they create presences where previously the eclipsing sounds of the victors had drowned out the voices of the defeated.

With the goal of describing and presenting Jamaica Kincaid's poetics as a discourse that performs impermanence through text, this study considers the author's major fictional works in addition to select shorter and nonfictional texts. The main part of my analyses focuses on the five novels *Annie John* (1985), *Lucy* (1990), *The Autobiography of My Mother* (1996), *Mr. Potter* (2002), and *See Now Then* (2013).[3] Susheila Nasta fittingly describes Kincaid's oeuvre as a family album, with each book attending to individual members of an assumed family unit. The earlier texts *Annie John* and *Lucy* – as well as the shorter texts in *At the Bottom of the River* (1978) and the uncollected text "Biography of a Dress" (1992) – detail the lives of girls growing up and away from their nuclear families. *The Autobiography of My Mother* focuses on a mother figure and female genealogy,[4] while *Mr. Potter* is interested in a father figure and explores a possible paternal genealogy. Finally, *See Now Then* details family life from a parental perspective.[5] What all of these texts have in common is their presentation of colonial history and ideology as pervading all aspects of life. All of the novels are set in the twentieth century and highlight the intense entanglement of collective history with matters that are intensely personal as well as mundane – be it family life in *See Now Then*, menstruation in *Annie John*, or the hobby of photography in *Lucy*.[6]

These texts centrally negotiate the relationships of past, present, and history. They simultaneously demonstrate the absence of representations of the past that

3 From here on, I will use the following abbreviations when quoting from Kincaid's works: *AJ* for *Annie John*, *ABM* for *The Autobiography of My Mother*, *MP* for *Mr. Potter*, and *SNT* for *See Now Then*.

4 Chronologically, in 1997 follows *My Brother*, Jamaica Kincaid's non-fictional account of the loss of her brother to AIDS.

5 If read autobiographically, as some scholars do, this family album could be assumed to portray Jamaica Kincaid's own family. However, as these texts are fictional, I am interested in reading them as interconnected parts of a body of works, but not as direct continuations of each other. Nonetheless, see for instance Leigh Gilmore's informative thoughts on seriality in Kincaid's works (*The Limits of Autobiography*).

6 A second topical strand in the body of Kincaid's works is her travel and garden writing, including her monographs *A Small Place* (1988), *My Garden (Book):* (1999), and *Among Flowers: A Walk in the Himalaya* (2005). These nonfictional texts (including *My Brother* (1997)) are hardly less complex in literary form, but as my main objective is to investigate the techniques of the fictional text in its engagements with history, my analyses refer to Kincaid's nonfictional works only on occasion.

would adequately relate to the various frameworks of the present in which the characters exist, and that critique the overbearing presence of representations that continuously frame the characters as inferior subjects to colonial ideals.[7] In this, the texts grapple with history as a discursive construct that limits contemporary lives. In their resistances to overbearing colonial history, Kincaid's works dispute representational sovereignties over the past. Édouard Glissant describes the eclipsing presence of colonial history as a History "with a capital *H.* It is a totality that excludes other histories that do not fit into that of the West" (*Caribbean Discourse* 75). With "totality" Glissant here denotes "the unprecedented ambition of creating man in the image of the Western ideal" (*Caribbean Discourse* 75), i. e., Western discourse's claims to universality, which negate all other voices. Totality for Glissant is characterized by the superimposition of fixity and immobility (Glissant, *Poetics* 14, 171) on others. In contrast, he explains that the

> Caribbean is the site of a history characterized by ruptures and that began with a brutal dislocation, the slave trade. Our historical consciousness could not be deposited gradually and continuously like sediment, as it were, as happened with those peoples who have frequently produced a totalitarian philosophy of history, for instance European peoples, but came together in the context of shock, contradiction, painful negation, and explosive forces. The dislocation of the continuum, and the inability of collective consciousness to absorb it all, characterize what I call a nonhistory. The negative effect of this nonhistory is therefore the erasure of collective memory. (*Caribbean Discourse* 61–62)

Glissant then calls on writers to face the shock of nonhistory. While literature has similarly functioned as an instrument of Western totality in that it equally epitomized the legitimacy of the West,[8] it also bears the possibility for change: "[T]he writer must contribute to its [the Caribbean's] tormented chronology [...]. Literature for us will not be divided into genres but will implicate all perspectives of the human science. These inherited categories must not in this matter be an obstacle to a daring new methodology where it responds to the needs of our situation" (Glissant, *Caribbean Discourse* 65). Glissant sees literature as capable of surmounting the impediment of nonhistory, the totality of established literary form, and the limits of scholarly fields. With the goal to speak to current needs, Glissant maintains, literature may produce new approaches to Caribbean past, present, and

7 Michael Dash concisely summarizes this problem in colonial contexts: "History ultimately emerges as a fantasy peculiar to the Western imagination in its pursuit of a discourse that legitimizes its power and condemns other cultures to the periphery" ("Introduction" xxix).

8 Glissant here refers to and generalizes the significance of William Shakespeare's *The Tempest*, which elevates Prospero and neglects Caliban. See esp. 71–77 in *Caribbean Discourse* on the longstanding relations of history and literature, which Glissant identifies in their common projection of Western totality.

the precarious relationship of the temporalities.[9] "The production of texts must also produce history, not in its capacity to facilitate some happening, but in its ability to raise a concealed world to the level of consciousness" (Glissant, *Caribbean Discourse* 107).[10] In this study, I show how Jamaica Kincaid's works seemingly respond to this call to the pen in creatively engaging with history and producing Caribbean voices and perspectives previously absent from history.

In a literary writing of history, Glissant calls for a poetics that breaks with linearity and fixity of Western discourse. With his chaos theory, he offers a counterstragy to fixation. For Glissant, chaos as a positive force means "the breakdown of systems of standardization and uniformity, of 'l'universel généralisant' that is produced by imperialistic ideologies. If chaos means the alternative to this homogenizing force, it does not signify incomprehensible turbulence," as Michael Dash explains (*Édouard Glissant* 177).[11] Glissant advocates a "poetics that is latent, open, multilingual in intention, directly in contact with anything possible" (*Poetics* 32).[12] I read and further explore such openness in Kincaid's texts with Henry Louis Gates's concept of "Signifyin(g)," which explicates the rhetorical strategies of "repetition with a signal difference" and "black double-voicedness" (Gates, *Signifying Monkey* 51). Gates starts out with examining the construction of the linguistic sign as it is disrupted by Signifyin(g): as a sign that empties "the signifier [...] of its received concepts and fill[s] this empty signifier with [its] own concepts. By doing so, by supplanting the received standard English concept associated by (white) convention with this particular signifier, [it] (un)wittingly disrupt[s] the nature of the sign=*signified/signifier* equation itself" (Gates, *Signifying Monkey* 46, em-

---

9 See the following chapter on structures of time for a discussion of how Glissant proposes literature may take up historical discourse. There, I develop this notion in more detail and with regard to Kincaid's *See Now Then.*

10 With what she calls "critical fabulation," Saidiya Hartman proposes a similar approach to filling the gaps of history. However, Hartman is more reliant on historical archives, which Glissant is not interested in at all. See my discussion of Hartman in chapter 2.1. Birgit Neumann's "fictions of memory" might also insert themselves in this field of absences in that they produce "imaginative (re)construction of the past in response to current needs" (334). However, I am not foremost interested in personal (or cultural) memory work per se, but in engagements with representations of history. The reflection of the workings of memory, which are part of Neumann's concept, are thus not of central interest to my study.

11 See more on Glissant's poetics of chaos in the following chapter on "Time."

12 In this interest in non-linear writing, Glissant of course is not alone. For instance, Caribbean thinkers Kamau Brathwaite (Tidalectics), Derek Walcott (the sea and its multidirectional movements as metaphor for history), and Antonio Benítez-Rojo (The Repeating Island) have postulated similar poetics. See Eddie Baugh, "Literary Theory and the Caribbean." Glissant, however, is especially useful in the context of my own study, as he lays bare the connection of literary form and colonial history.

phasis in original). In this, the signifier remains the same, but the signified is supplanted with the rhetorical strategies themselves (Gates, *Signifying Monkey* 47). Thus, "[d]irecting, or redirecting, attention from the semantic to the rhetorical level [...] allows us to bring repressed meanings of a word, the meanings that lie in wait on the paradigmatic axis of discourse to bear on the syntagmatic axis" (Gates, *Signifying Monkey* 58). Gates observes this on a tropological level as well as in intertextual relations.

As a fundamental thought concept, Signifyin(g) guides me in thinking about Kincaid's repetitions with differences, when her texts negotiate received representations of the past that are changed to bring to light repressed versions. With Gates, I am particularly interested in formal revisions of text and image that then produce revisions of their significations. Through repetition and redirection, I argue, Kincaid's works dissipate representational fixity – especially in binary structures, which Glissant identifies as a key feature of Western totality. Hence, my general theoretical approach draws on both Gates's and Glissant's work to understand how formal revision constitutes resistances to the History with a capital *H* in Kincaid's works.[13]

To describe Kincaid's poetics, I use the term "impermanence." Glissant emphasizes a resistance to fixity, which I also find in Kincaid's works in various textual movements that effect impermanence. I take the term "impermanence" from a 2020 essay by Kincaid, in which she employs the image of a snow dome to laud the possibility of change through impermanence. Shaking everything up makes it possible for slightly changed views on the landscape within the dome to emerge.[14] The present study shows how Kincaid's writings of history shake up perspectives on the past, again and again, by changing representations and reenacting the past in different ways. Text, images, and performative acts shake up history, just as the plastic snowflakes in the snow dome are given a good shake from time to time. With the "poetics of impermanence," I thus seek to describe the techniques that Kincaid's texts use to approximate and appropriate history. Kincaid's

---

13 Beyond this framework, in my individual analyses I operate with further theoretical angles as they pertain to the topics at hand. Among others Saidiya Hartman on critical fabulation (ch. 2.1), Renate Lachmann on intertextuality (ch. 2.2), Homi Bhabha on colonial mimicry (ch. 2.2), Bakhtin on heteroglossia (ch. 2.3), Roland Barthes, Susan Sontag, Harvey Young on photography (ch. 3), Mikhail Bakhtin and Julia Kristeva on the grotesque (ch. 4.1), Judith Butler on performativity (ch. 4.2), Anne McClintock on S/M (ch. 4.2). My discussions of these and other theoretical positions take place in my analytical chapters, where theory is directly applied to the results of my close readings. What all of these theoretical approaches have in common is their deconstruction of binarities through multiplication or transgression, which reflects my general interests in this study.

14 See the epilogue of this study for a detailed discussion of Kincaid's "Inside the American Snow Dome."

poetics of impermanence encompass a range of literary techniques and textual motions that operate similarly. Foremost, repetition with a difference in Gates's sense allows new meanings to emerge of what has already been written and read. Repetitions and reiterations create a back and forth between old and new meanings; nevertheless, this is no pendular movement between two fixed poles, as binaries are dismantled when the back-and-forth moves past starting points, reexamining and removing them in the process. This poetics thus redirects perspectives to previously unnoticed aspects. It performs and reenacts with differences or different actors to create presences in a history that previously was marked by absences. In these movements, the texts stage approximations and distancing towards individual topics and ideas that can at times remain arrested in the in-between. Yet in-betweenness here never signals stasis, but rather a refusal to decide. In a similar way, the lack of movement that can be found in some moments of the texts points paradoxically to the perpetual motions of Kincaid's poetics. Ultimately, all of the poetic motions I describe throughout this study perform a radical opposition to History with a capital *H* and the fixity of Western discourse as Glissant describes it. Kincaid's poetics of impermanence, I contend, decenters the stasis and totality of history and opens renditions of the past to ambiguity and multiplicity, by which different notions and perspectives on history are produced.

The conceptual approaches by Gates and Glissant and a focus on formal aspects have so far found little resonance in major studies of Kincaid, although scholarly attention to her works has certainly not been lacking since the 1980s. Since the mid-1990s, a dozen book-length publications have published on Jamaica Kincaid's works. Early studies (Simmons, *Jamaica Kincaid*, 1994; Ferguson, *Jamaica Kincaid: Where the Land Meets the Body*, 1994) accomplish important foundational work in opening up Kincaid's texts to an academic audience.[15] They largely explore the texts from the perspective of content and broadly sketch out important themes and fields of interest that were later taken up by other scholars, such as the mother-daughter relationships, treatments of colonial history, migration, feminism. My work draws from these broad-ranging scholarly foundations to focus in more closely on individual analyses based on detailed readings paired with select theoretical approaches in order to attend to the Kincaid's poetics.

Generally, book-length studies of Kincaid's works up to date explore two interlocking academic interests – the recurring and complicated mother-daughter rela-

15 See also Paravisini-Gebert, *Jamaica Kincaid: A Critical Companion* (1999), Edwards, *Understanding Jamaica Kincaid* (2007), or Snodgrass, *Jamaica Kincaid: A Literary Companion* (2008).

tionships pervading most of Kincaid's texts,[16] and their (auto)biographical aspects. This is also the case for the vast majority of shorter academic investigations of Kincaid's oeuvre. Indeed, many scholars read Kincaid's novels as thinly veiled autobiographies, and some even equate their protagonists with the real-life author.[17] Particularly Leigh Gilmore's and Jana Evans Braziel's work on autobiographical aspects is noteworthy in this context. Both scholars develop their observations on (auto)biographical features in Kincaid's texts into fruitful reflections on genre, and both propose to think about new categories of autobiographical writing. Braziel suggests "alterbiography" as a form of life writing that expresses collective concerns, while Gilmore challenges the idea of autobiographical completeness with her notion of "serial autobiography."[18] My own study differs from this body of secondary literature in that I am not interested in examining Kincaid's works for traces of the author's life or in ascribing the texts with existing or new genre categories.[19] Instead, I draw on questions of generic form only to interrogate the textual strategies in Kincaid's texts that perform subjectivities and thus withstand colonial silencing within the texts.[20] My focus is on what the texts *do*, not on what they *are* in terms of categorization. Nor am I interested in extensive readings of Kincaid's mother figures for their own sake. As mentioned, other scholars have sufficiently covered this area of research. Although I cannot and do not leave the mother(s) unmentioned in my own inquiries into treatments of colonial history

---

16 For instance Bouson, *Jamaica Kincaid: Writing Memory, Writing Back to the Mother* (2005), Brancato, *Mother and Motherland in Jamaica Kincaid* (2005), Dance, *In Search of Annie Drew. Jamaica Kincaid's Mother and Muse* (2016). Dance's monograph is especially noteworthy here. While other readings rather focus on the mother within Kincaid's works – often reading her allegorically for a mother-country (see for instance Paravisini-Gebert, Brancato, Ferguson), Dance sets out to recover the actual Ms. Annie Drew, i.e, the mother of the author, who evidently inspired the texts' fictional mother figures. Based on her meticulous research and interviews (esp. in Antigua), Dance paints a picture of Annie Drew and thus seeks to exonerate her from the less than charitable representations in Kincaid's fictional works. Dance's biographical project moreover exemplifies the typically colluded readings of (auto)biography and fiction in Kincaid's works in which the mother figure all too frequently is 'decoded' as the author's real mother.

17 For specific attentions to possible autobiographical aspects, see for instance Ferguson, Bouson, Donnell, Coppola, Cousineau, or Paquet.

18 See chapter 2.3 on genre in Kincaid's works for a detailed discussion of Gilmore's and Braziel's approaches to life writing.

19 Hence, a description of the author's life is not included with this introduction. As elements of Kincaid's biography are, however, relevant to her situatedness in literary canons, a contextualization of Kincaid's person and work is part of chapter 2.2 on intertextuality.

20 See chapter 2.3 "Writing between and across Genres." See also my analysis of the text-image relations in a play with authenticity in chapter 3.3 "Photographic Product(ion)s of History."

(especially for their representations of colonial ideals), I do see them as only one part of the power structures I expose in my analyses.

The two collections of articles edited by Harold Bloom and Linda Lang-Peralta, respectively, offer a representative overview of the range of topics generally covered by shorter analyses of Kincaid's works (besides the foci on mother figures and autobiographical dimensions), such as the texts' use of language in their resistances to imperialism, intertextuality, visuality, gender and feminism, identity and self-making, diaspora and migration, death and mourning, orphanhood, childhood, colonial history, and garden and travel writing.[21] Many journal articles and book chapters written about Kincaid are notable for their detailed analyses of less popular aspects of her works. Yet the limited space for their arguments means that their claims mostly remain uncontextualized within Kincaid's large body of work.[22]

Of the monographs on Kincaid, Jana Evans Braziel's is the only thematically specific book to date, and one can say that it transcends the mapping of the field that earlier monographs so usefully undertook. Braziel's *Caribbean Genesis: Jamaica Kincaid and the Writing of New World* (2009) explores the topics of genesis, genealogy, and genocide and their entanglements in world making in Kincaid's texts through the intertextual lens of the Bible (particularly the books Genesis and Revelations). In this, transformations of autobiographical elements are at the heart of Braziel's approach. Her work seeks to understand "the process of writing autobiographical truth from historical lies and colonial legacies. The writing of autobiography, then, is not a seamless, transparent process but rather an ambivalent venture into efforts at autogenesis and against the forces of historical genocide" (Braziel, *Caribbean Genesis* 7). Interested in a similar range of topics in Kincaid's oeuvre, I too explore the ambivalences Braziel finds in Kincaid's autobiographical negotiations of the annihilating forces of history. I understand what Braziel terms "genesis" and compellingly reads as a biblical speaking into being as the enactment of presences that I observe in the engagements with history performed in various media. In this, genre is but one area in which literary strategy and textual move-

21 See for instance the following articles on specific topics: Castro, Matos, or Réjouis on resistance through specific language use; Chansky, Bickford, Gervasio, Rodriguez, or Lima on intertextuality; Vasquez, Lyle, Nasta, Emery, or Hirsch on visuality; Narain, Mardorossian, or Mahlis on gender and feminism; McCormick, Adams, or Anker on identity; Forbes, Nichols, or Murdoch on migration; Brophy, Soto-Crespo, or West on death; Ball, Arbino, Mårdberg, or Everett on orphanhood; Braziel, Simmons, or Rampaul on childhood; Marquis, Journey, or Jay on colonial history; Pećić, Knepper, Azima, or Tiffin on gardening.

22 Particularly noteworthy for their innovative topics are for instance the analyses of cloth and clothing by Matos or Sharrad.

ment creates ambivalences, destabilizes colonial legacies, and produces new meanings.

Braziel observes the nexus of genre, genealogy, and genocide by chronologically attending to each of Kincaid's major works on its own, as do all other monographs on Kincaid. My own study, by contrast, is structured around thematic foci, namely the media text, image, and body through which Kincaid's works engage with colonial history. Reading the Kincaid's individual texts *with* each other rather than separately, my analyses thus take a more comprehensive perspective on the texts' literary techniques and Kincaid's poetics of impermanence.[23]

I observe such techniques of impermanence in three different media that engage with history in Kincaid's works. Accordingly, my analyses ask how each medium – text, image, and body – participates in the revision of renditions of the past that structures the main questions of the present study. Chapter 1 on "Time" directly follows up on the issues sketched out in this introduction. Demonstrating the entanglements of temporalities, the chapter presents the notion of temporal impermanence which is at the heart of the textual engagement with history in Kincaid's two most recent novels *Mr. Potter* and *See Now Then.* My analyses show how history plays a role in the present and how media – especially texts – constitute representations of the past. The intersection of representations and temporalities projects a sense of impermanence on the formal aspects of text. With this focus, the first chapter lays the foundations for my further thoughts in the following chapters on the representative and constitutive capabilities of media.

Chapter 2 on "Text: Narrative Encounters with History" explores how Kincaid's poetics of impermanence productively approaches history in text. Three subchapters attend to this on different textual levels. Chapter 2.1, "Writing Personal and Collective Memory," approaches repetition with a reading of *Mr. Potter.* Through repetitions of words or sentences, or the narrations of small events, the text changes referents and meanings and thereby creates possibilities of the past that had hitherto not existed. Chapter 2.2, "Writing with and through Other Texts," is interested in the same technique but applies it to the relationships of Kin-

23 Derek Walcott's famous dictum that in Kincaid's writing the sentence "heads towards its own contradiction" (Walcott in Garis 80) is often quoted, but rarely explored, which is where my own study sets in. See the mentions of the quote in the introductions of monographs by Bouson, Simmons, Dance. Here, it points to the complexity of Kincaid's writing, which baffles, but is rarely analyzed in detail. Similarly, Covi comments on Kincaid's strategy of repetition ("Jamaica Kincaid" 348–349), Simmons is interested in rhythm (*Jamaica Kincaid*), and Ferguson calls Kincaid's writing an "innovative format" (*Jamaica Kincaid* 2) that "presents discursive alternatives to a Western linear modality" (*Jamaica Kincaid* 163). I, here, follow up with an analysis of *how* Kincaid's textual strategies work.

caid's texts with other texts. This chapter shows how Kincaid's works intertextually engage with colonial thought and literary motifs. Through their rewritings, they challenge established meanings and the validity of colonial textual renditions of the past in general. The third chapter, 2.3, on "Writing between and across Genres" focuses on genre as a form of colonial fixation of text, particularly in categories of life writing. Here, I examine how Kincaid's *Autobiography of My Mother* and "Biography of a Dress" revise generic conventions and ask who can speak for whom in a postcolonial context. In this, as I will show, the narrators and characters displace colonizing discourse and establish their own voices.

Based on my considerations of Kincaid's poetics of impermanence in chapter 2, the subsequent chapter focuses on visuality and history. Throughout the author's oeuvre, images occur in different ways. Within the narratives, the characters Annie John and Lucy destroy and produce photographs, which changes representations of them and their surroundings from colonial framings to self-determined portrayals, as I show in chapter 3.1 on "Making and Unmaking Representations." The textual representation of visuality is heightened in Kincaid's use of ekphrases, which I analyze in chapter 3.2 on *My Garden (Book):*, which I read as a photo album of history. The interaction of the two modes of representation leads the text to critically representative capacities of media in general. Finally, in chapter 3.3, "Photographic Product(ion)s of History," I focus on the portraits that are included with two of Kincaid's novels, *Mr. Potter* and *The Autobiography of My Mother*, to explore how the books toy with common conceptions of photography in their intersections with fictional text and how the images thus lend their purported authenticity to Kincaid's fictional versions of the past.

Chapter 4 on "Bodies: Embodied Enactments of History" reads the human body as a medium of history. The chapter builds on my previous explorations of how text and image engage with colonial history, as the human body here is largely determined by colonial discourse and gazes that categorize and fixate. I examine how bodies are externally ascribed with meaning, and how they also performatively free themselves from colonial attributions and engage in self-making through bodily practices. The first subchapter, "Transgressions of Gender," reads female bodies as subverting expectations though grotesquerie (especially in *Annie John*) and through smelliness that upsets colonial order. It also explores sexuality as scripted in *Lucy* and *The Autobiography of My Mother*. In chapter 4.2, "Scripted Sexualities," I thus focus on concrete sexual practices through which female bodies negotiate colonial history. Through abortive practices, *The Autobiography of My Mother* emphasize the historical exploitation of female reproductivity during slavery. Yet the text also highlights the knowledge of abortifacients, thereby countering the objectification of the female body. I moreover read explicit sexual acts as stagings and subversions of colonial fantasy in S/M play. Reenactment here entails the reversal of received roles.

Ultimately, the entirety of my readings traces the enactments of self-determined presences in the representations of colonial history though these various media. Kincaid's works produce these representations by evaluating the representative capacities of media, thereby decentering established representations of history, and by asserting their own versions through ambiguity and the multiplication of perspectives.

1 Time: Temporal Impermanence and History

1.1 Perpetuity of the Present and the Presentness of the Past: Time and Temporalities in *Mr. Potter* and *See Now Then*

Time is ungraspable, as Reinhart Koselleck observes (*Zeitschichten* 305).[24] Accordingly, it can only be construed in the engagements with it and in language. Similar to history, concepts of time appear in discourse. The texts in Kincaid's oeuvre largely toy with chronology and linearity in one way or another. All of Jamaica Kincaid's writing is centrally concerned with time and temporalities: the narratives of individuation – *Annie John* and *Lucy* – intertwine colonial history and personal past, affecting the protagonists' processes of attaining selfhood; Kincaid's essayist garden writing treats the past and future seasons of the garden as equally important as the colonial legacy of transplantation; likewise the essayistic text *A Small Place* interrogates the Western tourist's neocolonial repetition of conquest in Antigua. The later novels *The Autobiography of My Mother* and *Mr. Potter* more overtly focus on ancestry and genealogies that reach back to the early days of colonialism, and Kincaid's latest book *See Now Then* contemplates not only the disintegration of its protagonist couple's marriage, but also the nature of time itself. Although all of Kincaid's texts are predominantly set in the twentieth century (roughly between the 1920s and the 2010s), they always entangle past and the present.

This opening chapter analyzes the concept of time and temporalities as found in the two novels *Mr. Potter* and *See Now Then* to understand the relationship of past and present in Kincaid's writings of history. In Kincaid's texts, temporalities overlap when the past affects the present, both with regard to the content of the narrative and the form of narration. Linearity is interrupted by a range of literary devices, such as when the frequent analepses of narrative repetition disrupt a clear chronological progression of the narratives (see chapter 2.1); circular narration lets the narrative end where it began (in *Lucy* and *See Now Then*); ekphrastic writing halts the narrative flow when attention to detail lets the duration of the discourse significantly exceed the duration of the story (see chapter 3.2); and intertextuality picks up on colonialist discourses, relating historic texts to contemporary ones (see chapter 2.2). Since the present study devotes separate chapters to each of these literary features, this first chapter elucidates the conception of time in Kincaid's works in more general terms through close readings of passages from *Mr. Potter* and *See Now Then.*

24 German original: “anschauungslos” (*Zeitschichten* 305).

 https://doi.org/10.1515/9783111027500-003

In his book on temporalities, Russell West-Pavlov notes that "temporal experimentation is emblematic of much postcolonial fiction, indeed of postcolonial thought in general, in the confrontation it stages with forms of temporality identified as specifically European and imperialist, and the ways it proposes alternatives that may elude this imperial time" (158). In the same way, Kincaid's works defy Western notions of time and thus engage with history. In these texts, history is not in the past; and the past is not what has passed, but is part of the present. The notion of time in Kincaid's works is characterized by a simultaneity that continuously approximates past and present to thereby demonstrate the lasting effects of colonial history.

Accumulation of Time: The One Moment of Five Centuries

"Time does not pass, it accumulates," Ian Baucom writes about time in the postcolonial condition (24).[25] This is also true of Kincaid's works: the past is not concluded and time piles up on top of it, resulting in extended nonsynchronous "nows." Roderick Potter, for instance, protagonist of the 2002 novel *Mr. Potter*, signifies the past within his present in the sense of Ernst Bloch's notion of nonsynchronism, defined as when "earlier bodies emerge in the Now and send a bit of prehistoric life into it" ("Nonsynchronism" 23). Paradoxically, however, Mr. Potter simultaneously lives in a perpetual present. The life story of Roderick Potter, a cab driver and chauffeur in modern-day Antigua, is invented by his daughter who has hardly met him. Exploring the reasons for his absence from her life, the daughter-narrator makes up the biography of a distant father figure as well as his ancestry. She determines that the legacies of colonialism and slavery in Antigua affected her paternal genealogy in such a way that it rendered her forebears unable to form intimate human relationships, which also explains Mr. Potter's neglect of his daugh-

25 While I appreciate Baucom's apposite choice of words and gladly take it up, I depart from his notion of time accumulating in that I differentiate between time, past, and history, where Baucom does not, as he subsumes all three in "time." To come to the conclusion that time accumulates, Baucom explicitly draws on a passage from *Poetics of Relation*, in which Édouard Glissant states that "[w]e no longer reveal totality within ourselves by lightning flashes. We approach it through the accumulation of sediments" (33, see Baucom 319). In my own reading of Glissant, what sediments is not time, but history, as I demonstrate in this chapter. Hence, when I take up Baucom's useful phrasing on time "accumulating," I specifically refer to time – in contrast to history.

ter.[26] In her literary examination of her father's life, the daughter-narrator thus also explores the relationship of past and present.

The starting point here for the ongoing past is the beginning of colonialism: "Mr. Potter's life time began in the year fourteen hundred and ninety-two but he was born on the seventh day of January, nineteen hundred and twenty-two" (*MP* 177). The year 1492 is marked by the arrival of Christopher Columbus in the Americas, an event that is commonly equated with the beginning of European exploitation of the continents' land and peoples. The coincidence of Mr. Potter's lifetime with the time of colonial conquest marks the protagonist as its very product. It also highlights that the last five centuries are one continuous time period and that the imperialist project still continues, since it keeps affecting a contemporary life.

With the "discovery" of the Americas, the year 1492 marks the inception of the "New World's" historiography. Interestingly, Mr. Potter's lifetime does not commence with the traumatic events of the Middle Passage, but with the events that established the Americas as an inhabitable space in Western imagination. This signals that Mr. Potter's life is determined not only by events in the past, but equally by their discursive distribution. Pointing to the significant year again, in her essay "In History" (in *My Garden (Book):*[27]) Kincaid asks: "Should it [history] be an idea, should it be an open wound with each breath I take in and expel healing and opening the wound again and again, over and over, and is this healing and opening a moment that began in 1492 and has yet to come to an end?" (*MGB* 153). Kincaid's question about the ontology of history points to some of its perceived characteristics: it began – just as Mr. Potter's lifetime – in 1492, which explicitly references the history of colonial conquest in the Americas (including the histories of those who were conquered, dislocated, extinguished). Secondly, history here is not perceived as a time span, but as *one* moment that is ongoing and has yet to conclude. The designation of history as a "wound" aligns it with trauma – a trauma that began in the year 1492. Again, as with Mr. Potter's lifetime, the year not only references events, but also the onset of a chronological historiography of the Americas so that these three characteristics equally apply to the discourses of history.

In both of the quoted passages from "In History" and *Mr. Potter,* the past in question is a colonial one, and it does not cease but constitutes a temporality of presentness that encompasses five centuries. Time here does not pass; it accumulates. Mr. Potter experiences the accumulation of time as a condition of stasis. To

---

26 See chapter 2.1, "Writing Personal and Collective Memory: Repetition in *Mr. Potter,*" for a detailed analysis of the daughter-narrator's textual production of her absentee father's life.

27 From here on, I will use the abbreviation *MGB* when quoting from *My Garden (Book):*

him every day is the same and this is at the center of his perception of the world (*MP* 178). The day as the preferred time unit highlights the aftereffects of colonialism and slavery on Mr. Potter, as he displays a perception of time that is based on that of a slave, whose only measure for time was the alternation of day and night (Glissant, *Caribbean Discourse* 84). The enslaved person here concurs with "the peasant, the agricultural worker" (Glissant, *Caribbean Discourse* 84), which emphasizes the focus on natural events, such as the sunrise and nightfall or the change of seasons, in this approach to measuring time. With Reinhart Koselleck, this understanding of time may be described as natural time (*Futures Past* 1, 95), which points to the simplicity of Roderick Potter's life.

Moreover, an expanse of time is inconceivable to Mr. Potter, so that after his seventy years of life, "all he had been seemed like a day, whatever that might be, a day" (*MP* 59). Unable to perceive a period of seven decades, he must again revert to the day as a time unit to think about himself. Hence, while Mr. Potter embodies five centuries of the colonial past, in his world time does not pass, but every moment stays the same – precisely because the past does not move on. In this, Mr. Potter's nonsynchronisms depart from Ernst Bloch's concept of contemporaneity of the noncontemporaneous ("Gleichzeitigkeit des Ungleichzeitigen"), as Roderick Potter is not a "remnant" of the past (Bloch, "Nonsynchronism" 31),[28] reemerging in the present to contradict it, but embodies the whole of the past since 1492.

Mr. Potter's nonsynchronous temporality is juxtaposed with another concept of time that is symbolized by a great clock on an Anglican church tower in Antigua. The tower's clock has four faces, each pointing to one of the "four corners of the earth" (*MP* 179). The location of the clock on a church's bell tower references two originating moments of the clock and of time concepts: The bell tower points to the etymological origins of the clock, which derived from the German "Glocke" or Flemish "clokke" (West-Pavlov 16). The bell's chime as an indicator of hours or quarters signifies the clock's central purpose as a chronometer and in communicating time over a distance, to which the four faces of the clock also attest, each one turned to a different cardinal direction. The church tower also indicates a Western linear notion of time. According to Johannes Fabian, this linear concept emerged from biblical chronology, in which "[t]ime was thought […] as a sequence of specific events that befall a chosen people" (2). In modernity, this gave way to linearity through secularization on the basis of generalization and universalization (Fabian 2). Michelle Wright reminds us of the link between linearity and Western narratives of progress. Moreover, "the linear progress narrative is what organizes most of our knowledge and knowledge production in the West," she writes (15–16).

28 German original: "ungleichzeitiger Rest" (Bloch, *Erbschaft* 116).

Mr. Potter, caught in repetitive small units of time, is excluded from progress and from attaining knowledge.[29]

The time indicated by the mechanical clock is moreover that of industrialization and capitalism (Koselleck, *Practice* 104). The connection of capitalism and time is highlighted by the cathedral having been built by enslaved Africans (*MP* 176), slavery being one of the epitomes of capitalist exploitation. In this way, the clock in *Mr. Potter* indicates that both colonial time (as in the historical period) and the colonizer's time (as in the concept of time) were constituted through the exploitation of others.[30]

In the eyes of the narrator, the cathedral "seem[s] as if it simultaneously captured and released time" (*MP* 176). Indeed, this is how a mechanical clock measures time. With its invention, time

> no longer flowed or trickled [as with hour glasses or water clocks], but was indexed, not embodied, by the clock, which stood for time itself. [...] The development of the escapement system, which translated the force of gravity into a series of stop-go movement, effectively divided time up into mechanical segments: lock, release, lock, release, audibly perceptible in the tick-tock rhythm. (West-Pavlov 17)

The narrator links this repetitive mechanical movement to Mr. Potter himself, who can only handle natural time units as small as days which are characterized by repetitive sameness. In analogy to the mechanical clock, Roderick Potter himself is described as "a definition of time captured and released, released and captured" (*MP* 176). The linking of time, clock, and Mr. Potter once more highlights colonialism's making of Mr. Potter: the clock as a time keeping tool signifies the time of the colonizer, while Mr. Potter signifies the effect of the colonizer on the colonized.

The ostensible oppositions of constant accumulation and repetitive stasis, of Mr. Potter's embodiment of a time span comprising five centuries while simultaneously living within the parameters of "disembodied" mechanical time (West-Pavlov 20), of presentness of the past and perpetual present(s), as exhibited in *Mr. Potter*, highlight the heterogeneity of simultaneous temporalities. Bloch attests different nonsynchronous temporalities to different social classes. Reinhart Koselleck expands this notion of the contemporaneity of the noncontemporaneous (*Futures Past* 95), which he describes as concurrent layers of time,[31] in differentiating between different kinds of time, such as natural time and historical time. While natural time is defined by an inherent chronology of naturally occurring events,

29 See chapter 2.2 on Mr. Potter's inability to reflect on the relationship of self and world.

30 See also, for instance, Gilroy or Baucom for studies of transatlantic slavery as the onset of Western modernity.

31 German original: "Zeitschichten" (*Zeitschichten* 9).

"[h]istorical temporalities follow a sequence different from the temporal rhythms given in nature" (Koselleck, *Futures Past* 96), i. e., historical time fundamentally differs from the progression of natural temporalities.

> Even the singularity of a unique historical time [...] can be cast in doubt. [...] [It] is bound up with social and political actions, with concretely acting and suffering human beings and their institutions and organizations. All these actions have definite, internalized forms of conduct, each with a particular temporal rhythm. (Koselleck, *Futures Past* 2)

Accordingly, Koselleck does not speak of *one* historical time, but of many times with their own temporalities that may exist simultaneously and are superimposed onto each other (*Futures Past* 2). Kincaid's Roderick Potter simultaneously represents natural time, mechanical time, and historical time when he lives from day to day while embodying centuries.

Achille Mbembe describes such superimposing and intersecting layers of time as the postcolony's "time of entanglement:"

> [It] is neither a linear time nor a simple sequence in which each moment effaces, annuls, and replaces those that preceded it, to the point where a single age exists within society. This time is not a series but an *interlocking* of presents, pasts, and futures that retain their depths of other presents, pasts, and futures, each age bearing, altering, and maintaining the previous ones. (16, emphasis in original)

Koselleck's and Mbembe's thoughts on time largely converge. However, Mbembe's "time of entanglement" determines the reasons for the interlocking temporalities in the postcolonial condition. Rather than identifying social conditions and political actions in general as a constituting factor of historical time, Mbembe points to the effects of colonial history (or histories) that cause the merging and intertwining of temporalities in the postcolony. In *Mr. Potter*, the entangled temporalities need to be thought together to make sense of his paradoxical signifying on a vast period of time while simultaneously remaining arrested in a heightened form of the present.

"Now, now, now": Time in Constant Flux and the Perpetuity of the Present

The sense of simultaneity and temporal entanglement that marks Mr. Potter is heightened further in Kincaid's later novel *See Now Then*. At first glance, the novel appears centrally occupied with the disintegration of the protagonist couple's marriage. Set entirely in their family home in Vermont at the end of the twentieth century, the text heavily draws on Greek mythology in examining the family life of Mr. and Mrs. Sweet and their two children Heracles and Persephone. Mrs. Sweet – protagonist, narrator, and fictional author – relates her thoughts and memories about her life, and while the family members' changing relationships are ostensibly at the center of her narrative, it is actually much more concerned with the nature of time, which plays a principal role in these relationships, as well informing the workings of Mrs. Sweet's memory.

See Now Then is replete with temporal levels that are neither clearly separated nor separable, as suggested by the book's programmatic title, which can be read in (at least) three different ways: 1. "see a then in the now," i.e., see the past in the present; 2. "see the now in a then that is yet to come," i.e., see the present in a future; and 3. "see the now from a position in a then that was," i.e., see the present in the past. Even with the three words that constitute the novel's title, the book conveys a plurality of meanings as well as temporalities.

The book's title is then reiterated in the first sentence of the novel: "See now then, the dear Mrs. Sweet who lived with her husband Mr. Sweet and their two children, the beautiful Persephone and the young Heracles in the Shirley Jackson house, which was in a small village in New England" (*SNT* 3). The verb "see" here is set as an imperative so that the novel's first sentence, like the title, appeals to the reader to imagine Mrs. Sweet and her family in his or her own present, which adds this extratextual temporality to the mix of temporal levels referenced in the title.[32]

After extending the invitation to "see" Mrs. Sweet, the text goes on to describe the house and Mrs. Sweet, who "just now" (*SNT* 4) is standing by one of its windows, beholding the sights before her, and letting her thoughts wander. The reader is thus introduced to the narrator and focalizer (of the most part) of the text, as what follows is an account of Mrs. Sweet's perceptions and memories, which convey her particular perceptions of time. The text's insistence on Mrs. Sweet's "now" indicates the individual position from which time is perceived, which recalls Michelle Wright's concept of "Epiphenomenal time." Wright proposes to understand historical constructs (of Blackness[33]) in conjunction with phenomenological man-

32 The phrase "see now then" repeatedly occurs throughout the book and in different combinations of the three words, which changes their meanings and allows them to refer to different temporalities in each reiteration. See for instance *SNT* 12, 30, 34, 65, 71.

33 Throughout this book, I capitalize "Black" to denote a position defined by discursive constructs, be it in historical, socio-economic, or political categories. While Blackness is part of the subject positions of the characters in Kincaid's texts, it rarely is at the center of the texts' concerns her body of works. Rather, the texts explore power positions, which are tied to Blackness as much as to other categories, such as personal relationships, education, gender, or class. In this, Kincaid's texts relate to Michelle Wright's reading the historical events to which Blackness often is tied (such as the Middle Passage) as epistemologies – "narratives of knowledge that are taught, 'facts' of Blackness"

ifestations of Blackness, which are denoted by "*Epiphenomenal time*, or the now, through which past, present, and future are always interpreted" (4, emphasis in original).[34] These current moments through and in which the past is perceived are also at the center of *See Now Then*'s narrative situation(s).[35] The moment by the window at the beginning of the novel then establishes a frame for the ensuing narrative, as at the end of the book, after the narration of circa thirty years of family life and its disintegration, which culminates in Mr. Sweet leaving his wife, Mrs. Sweet again – or still – stands by the window contemplating. This circularity of narration suggests that the action between page 1 and page 182 only take place in Mrs. Sweet's mind. The two window scenes thus set a basic temporal level from which the narrator's thoughts begin to migrate.

Beyond that it is hardly ever clear in which present or past the narrator and the characters are situated. Temporalities keep superseding each other throughout the narrative, as for instance in the following passage:

> But now, there, just outside as he [Mr. Sweet] looked out the window, was young Heracles saying, Dad, Dad, and young Heracles was playing golf now, [...] but then, years later, now, now, now, the young Heracles, when asked to look back on the wreckage that had been made of his young life [...] in the rearview mirror [...] replied, "Yes, but objects in the mirror are closer than they appear." (*SNT* 26–27)

In the course of this passage, the adverb "now" changes its referent with each iteration. Initially, in "[b]ut now," it refers to the present in which the action of the narrative and presumably also the utterance is set. In the second iteration, when Heracles plays golf, "now" refers to a present within the narrative that might briefly be set after the first one, and it changes in focalization so that the two presents differ in who perceives them. The following threefold repetition "now, now, now," refers to a present set years after the episode under the window, when Heracles is asked about his youth, and again it presumably also refers to the present in which the narrator articulates or writes these words. The reiteration

(Wright 7–8). The analyses in the following chapters show how Kincaid's poetics of impermanence deconstructs such knowledge with textual strategies and performances of difference.

34 Wright capitalizes her use of the term to distinguish it from "its formal philosophical definition, in which the epiphenomenal is not in itself causal but nonetheless correlates with causal phenomena. [...] 'Epiphenomenal' time denotes the current moment, a moment that is not directly borne out of another (i. e., causally created)" (4).

35 *See Now Then* emphasizes that time is perceived individually and that is centrally affects other perceptions. For instance, "time and space intermingle[e], becoming one thing, all in the mind of Mrs. Sweet" (*SNT* 6) and "the ways in which Mr. Sweet's imagination, his Now and his Then, his ways of seeing the present, the past and the future, colored the ways in which he saw her [Mrs. Sweet]" (*SNT* 88).

here might further be read as three different moments – three "nows" following one another immediately, which emphasizes the temporal level in which the narrator is positioned, just as it highlights the plurality of presents that make up time. Hence, in referring to a different point in time with each repetition, the adverb "now" demonstrates the entanglement of temporalities, as well as the ambiguous position of the narrator, who in her narrative moves through time in a nonlinear fashion. Moreover, the whole passage is set in simple past, which indicates that the action or situation took place in a completed period of time. Accordingly, the "nows" in the passage cannot be considered presents at all, but must also already be in the past.

The interlocking of pasts and presents is also emphasized by Heracles's smug response that objects in the mirror are closer than they appear. The mention of the rearview mirror – intended metaphorically in the question posed and taken up literally in the reply, when Heracles quotes the common phrase often printed on cars' rearview mirrors – suggests the necessity of a medium to perform retrospection, be it in a literal or in a figurative sense. The mirror allows vision to "bend back" and regard what is behind oneself. However, Heracles indicates that the act of looking back might be fallacious, as ultimately looking at what is either physically behind oneself or at what is in one's past is moved into the present by being looked at. In this way, the past is not in the past at all, but the mirror projects it into the present. In the case of the car's rearview mirror, the past – or that which (one) has *passed* – is projected to one's front, which is where the mirror is physically located. Media as tools to perceive the past are central to the narrator's engagement with time in *See Now Then,* as Mrs. Sweet's retrospection occurs in writing. This engagement with history via media is also at the heart of the following chapters, in which I observe the uses of text, image, and body in Kincaid's works.

Mrs. Sweet generally lives in a multitude of presents, although the present is not graspable as one entity since time is constituted by a mass of interlocking temporalities. As such, presents and pasts are inseparable (for instance *SNT* 6, 44, 90, 135). Whereas Mr. Potter lives in the repetitious time spans of days, Mrs. Sweet lives in perpetual yet changing nows. Even the past that occurs in her mind's eye is thus moved into the present. Although "Now is ongoing and never ceases," it is "always just out of reach" (*SNT* 167). Mrs. Sweet's time is made up of a perpetuity of presents which themselves may have been pasts before.

The intermingling of pasts and presents in *See Now Then* occurs when it comes to personal memory, such as in the passage discussed above, as well as with regard to more universal matters like geohistory or collective colonial past. For instance, Mrs. Sweet thinking about the landscape outside her kitchen window indicates a persistence of the past: mountains and valleys are "all that remained of a great geological upheaval, a Then that she was seeing Now" (*SNT* 10). The mountain range here still stands witness to the events of its creation and transformation, which are long past. Similarly, less material remnants of the past remain, when also events of human history keep affecting the contemporary lives of the protagonist couple.

> Mr. Sweet [...] grew up in the atmosphere of questions of life and death: the murder of millions of people in a short period of time who lived continents away from each other; on the other hand hovering over Mrs. Sweet, though she had been made to understand it as if it were a style of a skirt, or a style of a shape of a blouse, a collar, a sleeve, was a monstrosity, a distortion of human relationships: The Atlantic Slave Trade. What is the Atlantic? What is the slave trade? So asked Mr. Sweet. (*SNT* 12)

While the Euro-American husband did not live through the Holocaust, just as the Caribbean-immigrated wife did not grow up in times of slavery, the passage intimates that neither era is over yet, but that these "upheavals" (*SNT* 19) are as ever-present as New England's mountain ranges. All three upheavals – in a literal sense the geological formations, and in a metaphorical sense the Holocaust and the history of colonialism – persist in the present when the past does not pass and time accumulates instead,[36] just as the Earth's crust collected in the formations of mountains.

However, the accumulation of time and the perpetuity of the presents do not indicate that time remains static. As seen, the "nows" change frequently from one to the next and are interspersed with "thens." Time does not "pass" in a linear fashion, but it nevertheless is in constant flux: "the present will be now then and the past is now then and the future will be a now then, and [...] the past and the present and the future has no permanent present tense, has no certainty in regard to right now" (*SNT* 13). In Mrs. Sweet's narrative, time is in constant flow, made up of a multitude of intertwined different forms of time and temporalities, all of which can only be perceived in and through the present, which itself remains and infinitude of ungraspable moments.

Nonhistory and Times of Chaos. Or: The Chaos of Time

Mrs. Sweet and Mr. Potter are not "stuck in the past," but in an ongoing multitude of presents. The past rather "lasts into" the characters' presents than that they re-

36 Not only upheavals persist, but, as Mrs. Sweet surmises, positive events are generally excluded from public interest: "See Now Then, just to see anything at all, especially the present, was to always be inside a great world of disaster, catastrophe, and also joy and happiness, but these latter are not accounted for in history, they were and are relegated to personal memory" (*SNT* 65).

turn to it. Roderick Potter is unable to perceive the passing of time while simultaneously standing in for a whole history of colonialism. Mrs. Sweet lives in an entanglement of ungraspable "nows" and returning "thens" which keep moving into her presents. Such interplay of timelessness and the abundance of time situates both Mrs. Sweet and Mr. Potter in a condition of "nonhistory," as theorized by Édouard Glissant: "Our historical consciousness could not be deposited gradually and continuously like sediment [...] but came together in the context of shock, contraction, painful negation and explosive forces. The dislocation of the continuum, and the inability of collective consciousness to absorb it all, characterize what I call a nonhistory" (Glissant, *Caribbean Discourse* 61–62).[37] As the History with a capital *H* displaced other histories, their chronologies were lost and the sedimentation of historical consciousness became impossible. Nonhistory, as "[t]he implosion of Caribbean history [...] relieves us of the linear, hierarchical vision of a single History that would run its unique course" (Glissant, *Caribbean Discourse* 66). Now, Glissant continues, "time cannot be conceived as a basic dimension of human experience" (Glissant, *Caribbean Discourse* 84). This, however, does not mean that time or the past are absent. On the contrary, the past is "obsessively present," but "has not yet emerged as history" (Glissant, *Caribbean Discourse* 63).[38] In short, time keeps accumulating so that the past stays on into the present in a nonsynchronous simultaneity, precisely because the past has not yet become history.[39]

Kincaid's Mrs. Sweet seems to be caught in the same limbo of the past repeating itself when she muses that "in your own mind, you can see the series of events that are to come, that are arrayed before you, and they appear as if they are in the rearview mirror but only in reverse" (*SNT* 158). The instrument of retrospection, fallacious as it is, as Heracles points out, can only display further renditions

37 French original for "dislocation of the continuum": "discontinu dans le continu" (*Le discours* 223–224), and for "inability": "l'impossiblité" (*Le discours* 224). See the introduction of this study for a discussion of nonhistory and History with a capital *H*.

38 French original: "Le passé, notre passé subi, qui n'est pas encore histoire pour nous, est pourtant là (ici) qui nous lancine" (*Le discours* 226). Glissant here investigates the reasons for the absence of linearity and will later advocate the plurality of histories in *Poetics of Relation*. Glissant's phrasing that Caribbean history has *not yet* emerged (French: "n'est pas *encore*" – emphasis mine) seems to indicate a desire for linearity after all, as it implies that nonhistory could easily be overcome if only it were discursively treated in an appropriate fashion. However, since Glissant does not explore this further and since this remains implied only in the words "yet/encore," I do not follow this possible path of critique, but rather seek to pick up on the possibilities it offers to engage nonhistory that Glissant advocates himself in *Poetics of Relation*, namely chaos as an opportunity to counter the normativity of History with a capital *H*.

39 With "history," I here refer to a past that was thoroughly discursively treated and could thus be regarded as concluded in the sense of Glissant.

of the past. Mrs. Sweet here predicts the future in terms of Koselleck's "prognosis," which "implies a diagnosis which introduces the past into the future" (*Futures Past* 22). The rearview mirror can only produce visions of the past, projected into the future.

These positions on the prevailing of the past and the ensuing chaos of temporalities may also be assessed positively for their defiance of linearity. For instance, Wilson Harris asks: "What is timelessness? Timelessness surely means breaking fixed linear ruling patterns into nonlinear simultaneous movement of such patterns forwards and backwards. Such simultaneity brings us into the mystery of timelessness and helps the past to be re-creatively potent" (26). According to Harris, the convergence of past and present may then facilitate a productive engagement with the past instead of just accepting its traumatic effects. Nonlinearity thus defies imperialist policies. Similarly, M. Jacqui Alexander lauds the palimpsest of layered and interlocking of temporal levels: "The idea of the 'new' structured through the 'old' scrambled, palimpsestic character of time, both jettisons the truncated distance of linear time and dislodges the impulse for incommensurability, which the ideology of distance creates. It thus rescrambles the 'here and now' and the 'then and there' to a 'here and there' and a 'then and now'" (190). Alexander's palimpsestic notion of time withstands the overriding effects of linearity,[40] but more than that, its rescrambling of "here and now" with "then and there" suspends othering mechanisms based on the intertwined concepts of space and time.[41]

Wilson's and Alexander's notions of the chaos of time and temporalities celebrate the defiance of Western concepts of linear time and chronology, commending a restoration of agency to the postcolonial subject that resists linearity. Édouard Glissant and Jamaica Kincaid, I argue, take a step beyond that when they find even more productive ways to engage with the effects of nonhistory. Both Kincaid and Glissant find their solutions to the predicament of nonhistory and its scrambling of temporalities in writing, as I show in the following, last section of this chapter.

---

40 See also Michelle Wright, *The Physics of Blackness*. Wright argues that "Epiphenomenal time" challenges linear progression narratives, such as Western historiography, through a consideration of the current moment that affects the perception of historical constructs.

41 See for instance Johannes Fabian, *Time and the Other*, on conjunctions of othering and time concepts.

Texere: The Knitting of the Past

Jamaica Kincaid, her fictional author Mrs. Sweet,[42] and Édouard Glissant all employ writing to productively engage with the past. The two (or three) writers' strategies resemble each other in that for Glissant writing becomes a tool to challenge History with a capital *H* and to produce other options to approach history, and for Kincaid (and Mrs. Sweet as well as the narrator of *Mr. Potter*) text is a medium to produce multiple versions of history, which entails the possibility to also make changes to it. Writing the past produces temporalities in the first place, as only the engagement with the past lets it emerge as something that has knowingly passed. It is such poetic production of time, past, and history that bears the possibilities of decentering the totality of History and to destabilize its perceived omnipotence to define past, present, and future.

In this form of writing, chaos defies linearity, as Harris and Alexander see it, but it becomes a productive moment, as Kincaid's Mrs. Sweet makes clear when she notes that "the story of creation, [...] the nature of any story, the story [is] the definition of chaos, of the unstable, of the uncertain, of the pause that holds the possibility of nothingness" (*SNT* 148). Chaos here holds the chance of annihilation, but more than that, its very instability and nonlinearity simultaneously also promise creation.

In *Poetics of Relation*, Glissant sees the possibility of "breaking the clear linear order to which Western thought had imparted such brilliance" (71) in the plantation system, which shattered memory and time. The plantation's destructive force, in Glissant's view, has a positive side to it, as it also created relation between cultures in that it was a space in which cultures clashed and which thus brought forth multilingualism. In short, the plantation was a birthplace for nonhierarchical cultural *métissage* (Glissant, *Poetics* 74), which is at the heart of Glissant's concept of relation. Glissant's *Poetics of Relation* elaborates his "chaos theory," which is already anticipated in *Caribbean Discourse.* Chaos here has thoroughly positive connotations: "[T]he way Chaos itself goes around is the opposite of what is ordinarily understood by 'chaotic' and that it opens onto a new phenomenon: Relation, or totality in evolution,[43] whose order is continually in flux and whose disorder one can imagine forever" (Glissant, *Poetics* 133). The poetics Glissant advocates are charac-

42 Mrs. Sweet overtly presents herself as the author of *See Now Then*, when the narrator (Mrs. Sweet) notes about her protagonist (Mrs. Sweet) that "she writes this" (*SNT* 132). Deictically, "this" denotes the text in front of the writer as well as the one in the hands of the reader, which establishes Mrs. Sweet as the fictional author of *See Now Then.* The same applies to *Mr. Potter*, whose narrator also is the book's fictional author.

43 French original: "mouvement" (Glissant, *Poétique* 147).

terized by openness and rhizomatic structures, and are not only to be pursued in literature, but equally in human interactions. Similar to Alexander's notion of palimpsestic time, which disrupts ideologies of distance, in Glissant's view literary writing may accomplish the same: "The poet's word leads from periphery to periphery, and, yes, it reproduces the track of circular nomadism; that is, it makes every periphery into a center, furthermore, it abolishes the very notion of center and periphery" (*Poetics* 29). In Glissant's poetics of relation, writing – and notably literary writing – works as a countermeasure to the annihilating forces of the History with a capital *H* and its propagation of the singularity of the West, in that this poetics of chaos is open to nonlinear and hence nonhierarchical pluralities of histories, temporalities, cultures, languages, etc. In this, it resembles Kincaid's poetics of impermanence. The main part of this study shows how Kincaid's literary strategies of impermanence perform pluralizations of history.

Both Kincaid and Glissant use textile metaphors to relate to writing. Etymologically, "text" is a descendant of textile, since it stems from the Latin tex-êre, which literally translates as "that which is woven," "web," "texture" (see Barthes, "Theory" 39). Roland Barthes, too, describes text in textile terms:

> *Text* means *Tissue*; but whereas hitherto we have always taken this tissue as a product, a ready-made veil, behind which lies, more or less hidden, meaning (truth), we are now emphasizing, in the tissue, the generative idea that the text is made, is worked out in a perpetual interweaving; lost in this tissue – this texture – the subject unmakes himself, like a spider dissolving in the constructive secretions of its web. (*The Pleasure* 64, emphasis in original)

Barthes's notion of interwoven meanings and "quotations drawn from innumberable centres of culture" ("Death" 146) resonates with Glissant's poetics of relation, which conceives of text in a similar fashion, but goes on to abandon centers for a multitude of peripheries. Barthes's tissue of a text is woven of a variety of quotations from different contexts, by which Barthes discredits the "Author-God" and the notion that text might be "a line of words releasing a single theological meaning" (Barthes, "Death" 146), left there by the author for the reader to find. Glissant repeatedly uses the metaphor of weaving (French: "tramer") in his writing about relation and poetics of chaos, for instance when arguing for the rights to difference and opacity within relation: "Opacities can coexist and converge, weaving fabrics. To understand these truly one must focus on the texture of the weave and not on the nature of its components. [...] Thought of self and thought of other here become obsolete in their duality. Every Other is a citizen and no longer a barbarian. [...] This-here is the weave, and it weaves no boundaries" (*Poetics* 190).[44] The weave

44 Besides making the image of the weave productive, Glissant emphasizes the importance of form

and weaving here are images for the positive aspects of chaos, namely interconnectedness and inseparableness, which occur both in Glissant's concept of relation and the performance of it in his writing about it.[45] While Glissant's web of a text strongly resembles Barthes's tissue in terms of its constitution, Glissant additionally emphasizes not the multitude of intertextual sources, but the heterogeneous plurality and simultaneity of meanings that might be entailed. Difference is celebrated and merges with the known without being assimilated.

Writing as weaving to come to terms with the past and to create a positive outlook on the future similarly features in Kincaid's own textile metaphor of knitting in *See Now Then.* Knitting here signifies writing, which interweaves strands of time. The two activities have the very same functions throughout *See Now Then.* They act as media of retrospection and creation, which also eventually merge into one and the same, as only retrospection creates time and only through the creation of temporalities can one look at past, present, and future.

Throughout the text, Mrs. Sweet continually knits, makes, and mends socks, blankets, and tunics to dress her family in (see *SNT* 58, 92, 160, 163–164), but eventually the text reveals that knitting here is more than simply a traditionally female task of homemaking and of filling time. It is – like writing – a tool for introspection and retrospection. Mrs. Sweet resides in her mind's eye while knitting garments (*SNT* 154) and contemplates time itself (*SNT* 158). A garment finally represents her life's story, when she knits "the shroud of her past, her childhood, her life before that, her life as it was interred in all the people she was descended and ascended from" (*SNT* 165). In the same way that making the garment may create the narrative of her life and the narratives of her ancestors, text is a medium in which the history is created.[46] Sitting at her writing desk, Mrs. Sweet "came alive in all her tenses, then, now, then again" (*SNT* 20, see also 71). Granted, not all moments

over content. Taking my cue from this, my work in the following chapters understands Kincaid's writings of history through their formal aspects, such as circular or repetitive structures, rather than solely their contents.

45 Michael Dash observes that "[i]t is as if Glissant is deliberately avoiding the temptation of a conceptual linearity or progression for a process of amalgamation, accumulation and circularity" (*Édouard Glissant* 175). Betsy Wing, translator of *Poetics of Relation,* also notes on Glissant's performance of the poetics of relation: "For a country whose history is composed of ruptures, to accept this linearity would imply a continued blindness to its own crazed history, […] and acceptance of the Western European epistemological principles that claim this history as its destiny. The structure of *Poetics of Relation* is based more on associative principles than on any steady progress toward irrefutable proof; it is an enactment of its own poetics" (in Glissant, *Poetics* xii).

46 Cloth as a medium is not singular to *See Now Then.* See also the dress that communicates Madame LaBatte's desires to Xuela in *The Autobiography of My Mother* (see ch. 4.2) or clothing as signaling community in *Annie John* (see ch. 4.1).

in which Mrs. Sweet remembers the past or when it moves into her presents are mediated by writing. However, considering that as the fictional author she wrote the text at hand, all of the reflections on the past are indeed produced in writing, as that is how the narrative is available. For the reader, the text of *See Now Then* provides a way to perceive Mrs. Sweet's temporalities, her thens and nows.

Moreover, both knitting and writing are acts of creation. The poetic aspect of Mrs. Sweet's work is emphasized by her husband's resentment of it: Mr. Sweet, himself a composer of classical music, regards his wife's knitting as "very disrespectful, for the creation of a thing is superior to the creation of a person, so thought Mr. Sweet to himself" (*SNT* 62). As the juxtaposition of giving birth and other creative production evinces, Mr. Sweet sees the realm of creation as that of men.[47]

In parallel to his dislike for Mrs. Sweet's creative acts in knitting, Mr. Sweet equally disapproves of his wife's writing. Her sitting down to write makes him sigh, "for in truth, everyone, anyone, in the whole world would know that he was the true heir to the position of sitting at the desk and contemplating the blank mounds of sheets of paper" (*SNT* 20 – 21). Mr. Sweet again displays the notion that creation – here through writing – should be a male prerogative. More than that, his reaction also suggests racist sentiment that literacy is a white trait.

When Mr. Sweet disregards the female side of creation, and especially the creation of a person, i. e., pregnancy and giving birth, his part in it, i. e., providing the sperm to fertilize the egg, is devalued, as the text suggests that Mrs. Sweet might be able to accomplish that on her own by way of her creative power: in a small altercation with their adolescent son Heracles – literally, a "run-in" – Mr. Sweet loses his testicles. This might be read as a metaphor questioning of Mr. Sweet's virility: he is a meager man in comparison to his physically strong son. Mrs. Sweet then creates a new pair of testicles from a pair of knitted socks, "the heels [...] imitating that vulnerable sac of liquid and solid matter that had been Mr. Sweet's testicles" (*SNT* 47). Metaphorically, Mrs. Sweet here mends the relationship between father and son and soothes Mr. Sweet's wounded pride. To convey this, however, the text employs its metaphor of creation, which ultimately supersedes Mr. Sweet's ability

47 This recalls such misogynist positions as Sigmund Freud's. He claimed that "women have made few contributions to the discoveries and inventions in the history of civilization; there is, however, one technique which they may have invented – that of plaiting and weaving" (164). Freud cannot, however, grant even this female creative power and muses that women do not create, but imitate nature's pattern of genital hair growth, which conceals the female – and in Freud's view deformed – genital organs (Freud 164). *See Now Then*'s emphasis on the creative power of handicraft disputes this.

to physically create human beings and demonstrates Mrs. Sweet's abilities to create as exceeding those of her husband (as he had suspected himself).

Mrs. Sweet's power to create through handicraft takes up classical discourses of female creation. The ancient Greek goddess Moira, personification of fate, created lives by spinning her threads. While for the most part fate is only one goddess in Homer's writing, it was later represented by the triad of the Moirae Clotho, Lachesis, and Atropos in Hesiod's *Theogony.* Here, Clotho spins the thread, Lachesis assigns the fate, and Atropos represents the fate that cannot be avoided, i. e., death, in her cutting the the thread of life (Smith 1109–1110). Spinning fates, Moira, or the Moirae, created lives through handiwork, deciding issues of life and death. This aspect also led to Moira's being thought of as the goddess of time, as she would decide the length of a human's time on earth (Roscher 3086). Mrs. Sweet, too, engages with time and with the thread of life when she knits it into her life's story. She is not subordinate to a preordained fate, but she takes her life into her own hands (even literally within the metaphor), knitting and purling her life herself.

Another noteworthy figure who combines handicraft and creation is Anancy the spider in Caribbean folklore. Anancy stories report how the spider outwits others, like Brer Lion, Brer Tiger, or Brer Monkey. This may involve genesis stories which might provide reasons for instance for wasps stinging, lizards waging their tails, or children crying over nothing, when the stories conclude with "Is Anancy mek it" (Tiffin, "Metaphor" 19). A trickster figure but also a creator, Anancy spins webs out of his own substance (De Souza 347) like he spins stories. Anancy travelled to the Caribbean from West Africa with enslaved people via the Middle Passage and is invested in Caribbean history, having served as a figure of resilience as the meek spider tricked powerful opponents in folk stories during times of slavery (Tiffin, "Metaphor" 20–22). A provider of knowledge (De Souza 348) and common sense (Bennett 67) in folk tales, in recent Caribbean writing Anancy symbolizes the creative postcolonial artist "who must write against the grain of Western tradition" (James 115).[48] As is the case with Anancy, also Mrs. Sweet's knitting eventually reveals itself as a metaphor for acts of writing. In a warped way, Mrs. Sweet hints at this when she notes that "metaphors are the true realm of the creator" (*SNT* 63), and Mrs. Sweet herself creates, in knitting, a metaphor for writing, which she herself has created in writing.

It is the position at the writing desk that ultimately connects writing and knitting, as the verbs seem interchangeable when she sits "in that infernal room in

48 See Kathryn Sullivan Kruger, *Weaving the Word*, for the connections of weaving and narration in cultures all over the world, esp. 24–25. For an analysis of reinterpretations of Anancy in Caribbean literature (Salkey, Brathwaite, Harris), see Helen Tiffin, "The Metaphor of Anancy."

which she brought alive all that had made her, that room of the kitchen that had in it the desk Donald had made for her, and she knitted a garment for herself" (*SNT* 168). The particular desk mentioned was previously identified as Mrs. Sweet's writing desk (*SNT* 20, 130) so the activity of knitting here reveals the metaphor.

In Mrs. Sweet's garment of her life, knitting and writing particularly merge together into a common *texere* in their capacities to create. As pointed out, the garment she knits for herself is that of her life's story, which is the text at hand. Remarkably, she not only creates the garment – and with that her own past – but she may also remake it if necessary. She

> tried to sort out how she came to be herself, and unraveling various parts of the garment that had been her own life: the hem of this garment had become undone, [...] from time to time it made her trip over her own self and stumble and she fell down and scraped her knees and bruised her forehead and her elbows too, the hem needs to be mended thought poor Mrs. Sweet, the hem needs to be made more secure. (*SNT* 164–165)

As Mrs. Sweet uses *texere* to perceive the past, she simultaneously also creates it in that only her engagement with it constitutes it as history. That she moreover may unravel it, mend it so that it will suit her better, shows that her creative power, which draws on impermanence as a basic principle, also enables her to change the past and its impact on her current life. This power to create and to change is where the possibility of healing the wound of history lies that is suggested in Kincaid's *My Garden (Book):*. Mrs. Sweet does not accept a predestined fate, but like the Greek goddess Moira, she takes the thread of her life into her own hands. The poetic making of both time and the past allows Mrs. Sweet not to be dominated by the absence of history and the chaos of temporalities, but to make them according to her own needs, just as she may change (the garment of) her life story so that it might not hurt her so.

Ultimately, as in Glissant's poetics of relation, in *See Now Then* the chaos of temporalities, the making and remaking of pasts and present, provides the productive possibilities to come to terms with traumatic pasts as the chaos allows for creation and making according to one's own needs. This creative approach to the past and its use of media, which is evident in Mrs. Sweet's engagement with time, is at the heart of Kincaid's poetics of impermanence throughout her works. The possibility of impermanence is ingrained in the basic notion of what historical representations are, as presented by these texts. The opportunity of poetic creation thus lies in this very understanding of how media work.

2 Text: Narrative Encounters with the Past

At the heart of Kincaid's textual engagements with the past, I see forms of repetition and redirection that keep the texts in a state of seeming impermanence. I argue that Kincaid's poetics of impermanence multiplies perspectives, dismantles hierarchies, and breaks with notions of fixedness. Doing so especially in its considerations of the past, Kincaid's acts of writing decenter History with a capital *H* and allow for new and different versions of history to emerge. Édouard Glissant writes that in art repetition offers the possibility to tap into an otherwise imperceptible movement, which may allow for different perceptions and the expression of yet-unconscious but already existing perspectives:

> Repetition [...] is an acknowledged form of consciousness both here and elsewhere. Relentlessly resuming something you have already said. Consenting to an infinitesimal momentum, an addition perhaps unnoticed that stubbornly persists in your knowledge. The difficulty: to keep this growing pile of common places from ending up as dispirited grumbling – *may art provide!* The probability: that you come to the bottom of all confluences to mark more strongly your inspirations. (*Poetics* 45, emphasis mine)

Continuous motion through forms of repetition that constantly take up previously unnoticed aspects is, I argue, likewise central Jamaica Kincaid's writings of history. Ultimately, the texts evince fixation and linearity because of their very medium. Text is read from left to right and remains on the page unchanged. Yet, as I show, Kincaid's poetics challenge this when they come back to what was written before (be it within her own texts or colonial intertexts) to reevaluate, revise, and lay bare the contemporary effects of past writing. In Henry Louis Gates's definition of Signifyin(g), "[t]o revise the received sign (quotient) literally accounted for in the relation represented by *signified/signifier* at its most apparently denotative level is to critique the nature of (white) meaning itself, to challenge through a literal critique of the sign the meaning of meaning" (*Signifying Monkey* 47). The process of repetition with a difference thus is meaning-making in itself. In Kincaid's texts it allows contemporary texts to poetically examine History, to expose its silences, and to conceive different and possibly more suitable histories.

In what follows, I observe such structures of repetition and impermanence in Kincaid's texts in three chapters. "Writing Personal and Collective Memory" explores the different levels of repetitions of single words and the narrative repetition of a singular event, which creates a new relationship between the narrator and the past in *Mr. Potter.* "Writing with and through Other Texts" analyzes intertextual rewriting of received representations of history and the simultaneous production of new versions of history. The third chapter, "Writing between and across

 https://doi.org/10.1515/9783111027500-004

Genres," investigates the roles of generic categories in a negotiation of who has the voice and agency to produce versions of history.

2.1 Writing Personal and Collective Memory: Repetition in *Mr. Potter*

Mr. Potter is the narrative of a daughter writing the fictional life story of an absentee father, whom she possibly never met. Her story of him serves to explain his historically caused absence from her and from anyone's life, as he – because of the destructive effects of colonialism on the descendants of enslaved people in Antigua – is unable to maintain relationships with other people. Through imagination, the daughter-narrator renders her father's life in fiction. In these fictions, repetition produces various "possibilities of the real," as the novel's narrator calls her multiple versions of the past. As I show, the narrator-daughter Elaine Potter uses this technique to claim her unknown father figure as her ancestor, while also creating a new sense of self in relation to her father Roderick Potter. While Elaine Potter's memory and narrative ostensibly focus on the past, they do not stagnate in it: instead, the technique of repetition with a difference propels the text forward into a future in which the past plays an important role, but in which it can also be evaluated and reevaluated according to present needs.[49] In this way, Elaine Potter writes her own personal history, tied in with colonial history, as Mr. Potter is a product of both a colonial past and colonial historiography that rendered him unnoteworthy to his daughter's writerly attention.

Writing (about) Roderick Potter, the narrator also produces somewhat fictional memories of Mr. Potter, as I show in the first section of this chapter on "The Creation of Memory." The second section, "Saying it Twice," shows how the narrator-writer uses reiteration to call versions of Mr. Potter into her text, through which she establishes a familial relationship with him. Employing repeated incantatory phrases, the narrator-daughter invokes a father who had hitherto hardly existed in her life. In a third section on "Meeting Mr. Potter," I examine repetition with a difference as a textual strategy that creates and elaborates an encounter of the daughter with Mr. Potter. Repetition here poetically produces a memory of a presence of an absent father and lets the narrator emerge as a self-made subject who is able to determine her own history.

49 While the technique occurs throughout the author's oeuvre, the 2002 novel presents particularly conspicuous cases of it, which even led one critic to deem it "plotless" (Roe). Because of this extensive use of repetition, the novel lends itself to an exemplary macrostructural reading of the technique.

The Creation of Memory: Imagination, Narration, Textualization

Elaine Cynthia Potter, the daughter of Mr. Roderick Potter and the narrator of *Mr. Potter*, presents herself as the (fictional) author of the book (for instance *MP* 87, 94, 185, 191).[50] She hardly knows her biological father since her parents separated before she was born. The daughter-author is aware that she produces her father – or rather a father figure – in fiction, which is as much informed by the narrations of others as by her own imagination. Both are unreliable. Fragments of the narrative of her father's life come to Elaine Potter from her mother: "And all this my mother has told me, all of this my mother has told me, my entire life as I live it is all my mother told me" (*MP* 127), the narrator explains, and thus points to the key element of *telling* in order to "make" a life. As the narrator regards her own life as made by her mother's narration, so she tells the life of her father into existence.

Facticity and reliability cannot be essential features of this form of biographical narration. The narrator excludes them from her source when she concedes, "I do not remember any of this, it is only that my mother has told me so and my mother's tongue and the words that flow from it cannot be relied upon" (*MP* 154). Further, Elaine Potter clearly states: "I am imagining this" (*MP* 54), evincing an awareness for the fictional nature of her memory work. She also emphasizes that her remembering happens in and through text: "These are all words, all of them, these words are my own" (*MP* 48), she writes, thereby claiming authorship for her father's and her own story instead of leaving the authorial power to the life-narrating mother.

Yet, while imagined, Elaine Potter's narrative of her paternal family members is still linked to past circumstances that must necessarily have existed, such as a biological grandmother. Of Roderick Potter's mother, Elfrida Robinson, Elaine Potter writes that, "she was of no account really, only she was the mother of my father and I cannot make myself forget that" (*MP* 73). The present-day narrator could hardly *remember* the details of Elfrida Robinson's life she relates, as Roderick Potter was absent from her life and could not have transmitted the knowledge of Elfrida's particularities. Moreover, as their later descendant imagines, Roderick Potter also suffered parental absence when his mother committed suicide while he was still an infant (*MP* 76). The particularities of her person and life can only be fictional. Writing Elfrida's life nonetheless lets the narrator-author access a link to the absentee grandmother. After Elfrida's death, as Elaine Potter writes, "no

50 Clearly, Kincaid here plays with the conventions of autobiography, as her own given name was Elaine Potter Richardson before she changed it to Jamaica Kincaid in 1973. Nevertheless, *Mr. Potter* is designated as a novel on the cover of the book and I here read it as such. Hence, when concerning the narrator and the author, I here generally refer to the fictional author of *Mr. Potter*.

one thought of her again [...] and only I now do so, think of her, and she was Mr. Potter's mother, my father's name was Mr. Potter" (*MP* 77). Simply thinking of that unknown person's (former) existence provides the granddaughter with a role in relation to Elfrida – she is the one who remembers.

Elaine Potter moreover never witnessed her parents' relationship. Here, she also construes thoughts which she then regards as memories:

> And in my mind, I turn over Mr. Potter and Annie Victoria Richardson [the narrator's mother], and they are in my memory, though that does seem an impossibility, that I could have known them before I was born of the two of them, and yet it is so: I have in my mind a memory of them from before the time they became my mother and father. (*MP* 137)

This quote is enlightening with regard to the narrator's conception of memory. She questions the possibility of "remembering" something she has not witnessed herself, which points to a notion of "memory" as a mental remnant of a past experience. "Experience" here would be defined by the physical and mental presence at an event and a form of participation in it. Still, Elaine Potter insists on the existence and validity of her own memory, as she emphasizes, "it is so." This conflicting relationship of the memory's existence and its impossibility evinces that in the understanding of Mr. Potter's narrator "memory" can indeed mean both the knowledge of a past personal experience and the imagination of a past event. While the two notions appear to mutually exclude each other, Elaine Potter accepts them as equivalent – especially in her project of writing her unknown relatives' lives.

Marianne Hirsch's concept of "postmemory" usefully speaks to the relationship of fiction and memory. "Postmemory" seeks to understand the generational transmission of memories of a traumatic past.[51] Hirsch describes this form of belated memory as

> the relationship that the "generation after" bears to the personal, collective, and cultural trauma of those who came before – experiences that they "remember" only by means of stories, images, and behaviors among which they grew up. But these experiences were transmitted to them so deeply and affectively as to *seem* to constitute memories in their own right. Postmemory's connection to the past is thus not actually mediated by recall but by imaginative investment, projection and creation. (*Generation of Postmemory* 5, emphasis in original)

---

51 Hirsch focuses on the Jewish Holocaust in Europe, but she concedes that the concept of postmemory "may usefully describe other second-generation memories of cultural or collective traumatic events and experiences" (Hirsch, *Family Frames* 22).

Elaine Potter's memories of her father are in part mediated by her mother's narrations of him and in part by her silences.[52] The mother thus filters her daughter's access to her biological father. His absence is emphasized by her silence, and his fictional presence is underscored by her unreliable narration. Nonetheless, the daughter-narrator claims her story of her father as her own, which highlights her own investment in the fictional making of a father figure. Taking her cues from necessary events in the past (such as births and deaths), she fictionally posits possible past events that serve as an explanatory model for the familial detachment.[53] The narrator-writer calls her versions of the past "possibilities of the real" (*MP* 137). Fiction here offers a form of access to the past, a way to establish relationships with people long dead, by telling versions of their lives. These might not be factual, but they allow their present-day relative to replace their absences with imaginations that establish and legitimize her as their descendant, and position her as "the central figure in Mr. Potter's life as he had been in [hers]" (*MP* 153). As such, facticity is not at stake here at all. Elaine Potter's narrative project replaces the void of parental absence with a fictional presence that consists of "possibilities."

Mr. Potter clearly is the central figure in the novel, not least indicated by the eponymous title of the book. The protagonist's name inescapably echoes through the pages, when the text foremost identifies Mr. Potter by the combination of the honorific "Mister," in the abbreviated written form of "Mr.," with his family name. Hence the name "Mr. Potter" occurs exceedingly frequent throughout the novel. Conspicuously, possible other denominations, such as "the man," "the chauffeur," "the Antiguan," are not used to identify the book's protagonist (except "father"). He is rarely called by his given name "Roderick," (for instance *MP* 48, 64, 98), and if so, most often still in combination with his family name. Less often even, Roderick Potter is called by an abbreviation of his first name, "Drickie," and if so, rather when referring to him as an infant or young boy (for instance *MP* 81, 89). Elaine Potter purposely refrains from calling her father "Drickie," as is made plain by her comment that "Drickie was the name people who loved

---

52 Elsewhere, *Mr. Potter*'s narrator declares: "And how amazed I was to hear my mother, who was alive, tell me that my father had died, for he was dead, she had never told me of his being alive" (*MP* 186).

53 The daughter-narrator finds such reasons in the ongoing effects of colonialism and slavery on family structures. Throughout her text, Elaine Potter presents her paternal family, especially Roderick Potter, as paradigmatic figures for descendants from enslaved people in Antigua. In that, she not only engages with her family history and her own experience of parental neglect, but also with the collective history of colonialism in the Caribbean.

him and knew him very well called him" (*MP* 153). Her choice of address as "Mr. Potter" positions her outside of this group of people.

The daughter-narrator presents her insistence on "Mr. Potter" as her protagonist's name as a conscious choice. She explains that "I came to know him by that name, Mr. Potter, and the name by which I know him is the way he will forever be known, for I am the one who can write the narrative of his life" (*MP* 87). The project of transposing an unknown person into the realm of knowability, even if only by name and not by individual particularities, is reminiscent of Saidiya Hartman's recovery project in "Venus in Two Acts." Lamenting the absence of narratives of the lives annihilated by slavery, Hartman seeks a way to "paint as full a picture of the lives of the captives as possible" (11). Not only are the lives of the enslaved destroyed, but slavery's archives equally perpetuate this violence in their dehumanizing records (Hartman 6). Drawing on Michel Foucault's intended project in "Lives of Infamous Men,"[54] Hartman ponders how to rescue life narratives from the colonial archive which "in this case, [is] a death sentence, a tomb" (Hartman 2) as it hardly transmits any knowledge, but only points to its absence. Finally, Hartman proposes what she calls "critical fabulation" (11): the voids of the archive, which denote gaps of knowledge may be bridged by imaginative narration. At the same time, these gaps cannot and should not be filled in, Hartman advises, "to respect black noise – the shrieks, the moans, the nonsense, and the opacity" (12). Ultimately, critical fabulation "is an impossible writing which attempts to say that which resists being said [...]. It is a history of unrecoverable past; it is a narrative of what might have been or could have been; it is a history written with and against the archive" (Hartman 12).

Like Saidiya Hartman, Elaine Potter has access to only fragments of Roderick Potter's life, but nonetheless seeks knowledge of him. The two projects differ in that Hartman is reliant on the archives of slavery, while Elaine Potter is dependent on the oral narratives of her mother and her willful unreliability. Elaine Potter's interest moreover is directed at her personal genealogy and history, while Hartman intervenes in the collective memory of transatlantic slavery.[55] The projects

54 Foucault is fascinated with the fragments of lives of people who have "been nothing in history, [...] having left no identifiable trace around them, they don't have and never will have any existence outside the precarious domicile of these words" (Foucault, "Infamous Men" 162). Foucault here refers to a collection of juridical documents denouncing people in abject positions. His "Lives of Infamous Men" was intended as the introduction to a recovery of the infamous from the archive.

55 However, as I argue in other chapters (1, 2.2), Mr. Potter is a paradigmatic figure who embodies the effects of colonial history. In this, Elaine Potter's memory work indeed takes up Caribbean colonial history beyond her personal family relations.

are nevertheless comparable in that both take up narration to negotiate absences. Although the past cannot be recovered, in *Mr. Potter* it is replaced with “possibilities of the real,” while Hartman looks for a “true […] picture” (13), mourns its impossibility, and finds her own writing “unable to exceed the limits of the sayable dictated by the archive” (12). I argue that Kincaid’s fictional author Elaine Potter is not held back by the same constraints. With her interest in “possibilities” of imagination, Elaine Potter’s project is more closely related to Tavia Nyong’o’s “afro-fabulation”: “A fabulist is a teller of tales, but he or she also discloses the powers of the false to create new possibilities” (Nyong’o 71). Mr. Potter’s narrator does not disguise her imaginative work, which accomplishes to produce versions of the pasts. “In its very spur toward the inauthentic, the not yet, the people who are missing, fabulation is always seeking to cobble something together, to produce connections and relations however much the resultant seams show” (Nyong’o 77). Similarly, Elaine Potter imaginatively transgresses the boundaries of knowledge set by her mother and expands it in her own repetitions and revisions, disregarding any external limits to her desire to understand and participate in history.

After having been informed of her father’s death, Elaine Potter travels to Antigua to visit his grave. However, the record of his burial site turns out to be lost and the grave master can recall only anecdotes about a fight erupting at the funeral, but not the location of Mr. Potter’s grave. Physical traces of Roderick Potter are lost and narratives are all that remains of him. Elaine Potter’s search for her father’s gravesite offers up a comparison with the project of yet another writer: In 1973, Alice Walker was on a quest for Zora Neale Hurston’s grave in Florida. After navigating several contradictory oral accounts of the last days of Hurston’s life, Alice Walker finds herself in graveyard with a local guide who is even less knowledgeable than she is. In spite of adverse circumstances, such as heat, waist-high weeds, biting bugs, and the risk of encountering poisonous snakes, Walker searches the abandoned cemetery in Fort Pierce, Florida, only to find nothing more than a small area in which the earth is slightly more sunken in. Not marked, it is solely a “spot that resembles a grave” (Walker 105). Walker admits, “to tell the truth, I can’t be positive that what I found is the grave” (112). Still, undeterredly Walker has a headstone erected at that very spot and claims having found Hurston’s grave (Walker 115), the site of which is now commonly acknowledged. In effect, Walker claimed Hurston’s grave, which could not be found anymore, just as Elaine Potter claims her father, who is not accessible anymore. Instead of a head stone, however, Elaine Potter commemorates Mr. Potter with a text, which presents a fictional version of his life. Faced with absence and inaccessibility, both women thus produce presences that hitherto had not existed, but are needed in their contemporary lives.

Saying it Twice: Incantations of Mr. Potter

In Elaine Potter's writing of her father's life, repetition is a key tool to create this father figure. The conspicuous repetition of the protagonist's name calls the character from obscurity into the text.[56] More so, it also assigns him and his narrator particular roles. The homodiegetic position of *Mr. Potter*'s narrator is not clearly evident from the text's outset. Before revealing her relationship with her protagonist, Elaine Potter is an omniscient heterodiegetic narrative instance relating a day in the life of a seemingly random character living in Antigua, who works as a chauffeur. More context is provided when well into the first chapter, the narrator introduces herself as the daughter of the text's protagonist (*MP* 20). The daughter-narrator thus textually slowly advances Mr. Potter. In the following chapters, the mentions of her key character's name are not only noticeable for their frequency, but also for the redundancy of their pairings with the declaration that Mr. Potter is Elaine Potter's father. The conjunction of name and role occurs at least twenty times throughout the novel's 195 pages,[57] often in appositional insertions, separated by two commas: "Mr. Potter, my father," (for instance *MP* 80, 86, or 157).[58]

With the same arrangements of vowels and consonants, both ending on "-er," the two key words conjoined in the phrase – "Potter" and "father" – are not only linked by their seemingly redundant pointing to the same referent, but also by their homeoteleutonic sound quality. This gives the many mentions of "Mr. Potter" a rhythmical quality, which cumulates in the threefold doubling of the incantation:

---

56 Victoria Bridges-Moussaron reads *Mr. Potter* as a prosopopoeia of the narrator's dead father, composed of rhythm and sound. Jana Evans Braziel emphasizes the aspect of genesis in *Mr. Potter*; see, e. g., her article "'Another line was born…': Genesis, Genealogy, and Genre in Jamaica Kincaid's *Mr. Potter*."

57 See *MP*, for instance 36, 48, 49, 50, 55, 57, 58, 64, 77, 80, 86, 100, 147, 153, 157, 185, 194, or 195. Indeed, one reviewer smugly notes that, "[i]n *Mr Potter*, Jamaica Kincaid seems to believe that if something's worth saying, it's worth saying twice" (Mars-Jones).

58 This certainly is not the only noteworthy stylistic device in Kincaid's texts in general and in *Mr. Potter* in particular. Here, I am only concerned with repetition as a means to create history. Most interesting, however, are also Gayatri Spivak's analysis of parataxes in *Lucy* (Spivak proposes that "with this literary characteristic of placing side by side without conjunctions, *Lucy* resists and alters any reading that would categorize it only by its subject matter," 338), and Nicole Matos's inquiry into the same device in *Mr. Potter* (building on Spivak's arguments, Matos suggests to regard the parataxes here as "withholding […] causal connections [and] as an ingenious, multi-functional ploy that engages questions of history, causality and genealogy in quite meaningful and consequential ways," in "Vicinity" 85).

Mr. Potter was my father, my father's name was Roderick Potter. (MP 57)

Mr. Potter was my father, my father's name was Roderick Nathaniel Potter. (*MP* 100)

Mr. Potter was my father, my father's name was Mr. Potter. (*MP* 195)

Separated and connected by the commas centered between them, the six clauses point to name and role of Mr. Potter in relation to his daughter. The commas ostensibly function as a mirrors between the sentences, the first one starting with the name, the second one ending with it, and the repetition of "my father" contracted in the center left and right of the comma.[59] However, since both name and the term "father" refer to the same entity, as the repetitions seem to enforce, all six paratactic clauses could be read as grammatically parallel, since all elements are set in nominative forms, which gives the sentences a chiastic structure. This back-and-forth movement in and between the three repetitions of the sentences in question both emphasizes and coalesces name and role of Mr. Potter, the father.

Writing about his concept of Signifyin(g), Henry Louis Gates notes that "[t]hinking about [it] is a bit like stumbling unaware into a hall or mirrors: the sign itself appears to be doubled, at the very least, and (re)doubled upon ever closer examination. It is not the sign itself, however, which has multiplied. [...] [O]nly the signifier has been doubled and (re)doubled" (*Signifying Monkey* 44). This disrupts "the nature of the sign=*signifier/signified*" (*Signifying Monkey* 46). In the case of the three sentences above, the signifiers are not doubled within the individual sentences. Yet all of them refer to the same signified, namely a thought concept of the entity of Mr. Potter. The name Mr. Potter and the term "my father" denote the same person. Then the phrase "my father's name" refers to the name of Mr. Potter, which underlines the component of the name as a signified, but which ultimately also indicates the entity of Mr. Potter. Finally, the name is mentioned again, now with the added given name(s). In their particular arrangement, four signifiers thus conflate the name and the role ("father"), two signifiers, which in themselves already point to the same thought concept – the personage of the father Mr. Potter. In her book chapter "Rhythm and Repetition: Kincaid's Incantatory List," Diane Simmons contends that "Kincaid's rhythms and repetitions, the long, seemingly artless, listlike sentences charm and lull the reader, disguising what is actually happening, which is that one thing is being transformed into another"

59 Bridges-Moussaron reads the phrases as "mirror image[s] of the other half, except for the addition of the word 'name' with which she gives him [Mr. Potter] what he denied her" (par. 17). I slightly disagree, as his name indeed was ceded to her (*MP* 153–154), what is lacking is the knowledge of the person from whom she inherited the family name.

(47).[60] The transformation here would be from name ("Mr. Potter") to role ("father"), i.e. from unacquainted man, known only by his last name, to a familial relation, and back to a person known by name, which changes the concept of "Mr. Potter." In the sense of Signifyin(g), the "[d]irecting, or redirecting, attention from the semantic to the rhetorical level [...] allows us to bring the repressed meanings of a word, the meanings that lie in wait on the paradigmatic axis of discourse, to bear on the syntagmatic axis" (Gates, *Signifying Monkey* 58) in this case of repetition in *Mr. Potter.* In the final mention of the name in the third example, which conspicuously is also the very last sentence of the book, the signified of the signifier "Mr. Potter" is transformed from being solely the name of an unknown person to the name of the person assigned by Elaine Potter. While the signifier remains the same, it now refers to a man who was appropriated by the writing of his daughter.

Quite obviously, moreover, the three sentences are not identical repetitions of each other, but they differ in that the name changes at the end of each sentence. The first sentence moves from honorific plus family name to the mention of given name with a repetition of the family name. Considering that "Mr. Potter" is the name the narrator-daughter chose to identify her father by (and by which she came to know him), it is not surprising that it would provide a starting point for her to claim him as a relative. The given name Roderick might signal somewhat more familiarity than the honorific "Mr.," which would point to an increased or intended closeness between the narrator and Mr. Potter. Moreover, the sentence is set in a context of exploring Roderick Potter's life. The addition of the middle name Nathaniel in the second sentence towards the center of the book establishes a relationship between the narrator-daughter and both her father and her grandfather from whom Roderick inherited his middle name Nathaniel. Claiming Mr. Potter as her father also lets her trace her paternal ancestry through names. Ultimately, however, the narrator reverts back to her initial choice of "Mr. Potter," which highlights the distance between father and daughter, as well as it finalizes the narrator's stipulation to make Mr. Potter known by *her* choice of name (see above and *MP* 87), for these are the very last words of the novel.

The family name "Potter," moreover refers not only to one particular man, but it is also the narrator's own name. However, it is not one she easily identifies with, as she explains: "[T]hat name, Potter, haunted me when I was a child, for I did not

60 In her analyses, Simmons rather focuses on lists of domestic objects and duties and provides examples from the short story "Girl," in which nurture is replaced with manipulation (*Jamaica Kincaid* 48). In *Annie John*, Simmons argues, the blissful union of mother and daughter is transformed into a painful separation (*Jamaica Kincaid* 49–50) and in *A Small Place*, the repeated phrase "you are a tourist" progresses from description to indictment (*Jamaica Kincaid* 52–54).

know any Potters […], and that name, Potter, was a part of my own name and yet I had never met the man whose name I bore" (*MP* 153–154). Hence the daughter's insistence on calling her father and the protagonist of her narrative by his last name also inscribes her own name into her text. The distance she creates by using the honorific address thus also implicates a proximity of father and daughter, indicated by the shared family name.

Elaine Potter plays out her claims on her father in text. In doing so, she seizes the power to know him, which his absence had denied her throughout her life. While her project cannot be one of restoration, as Mr. Potter is lost to her after his death in terms of possibly establishing a personal familial relationship, she can – through text – nevertheless create a version of him and in that "birth" a father for herself. Now she is the maker and he is removed to the powerless position she had previously inhabited by his doing. Her creation of her own father is explicitly accomplished in text: "And I now say, 'Mr. Potter,' but as I say his name, I am reading it also, and so to say his name and to imagine his life at the same time makes him whole and complete, not singular and fragmented, and this is because he is dead and beyond reading and writing and beyond contesting my authority to render him in my own image" (*MP* 193). The daughter concedes that her voice renders him mute,[61] which empowers her to create him on her own terms and in her own style of expression.[62]

With incantatory repetitions of her father's name, the daughter-narrator calls a man lost to her, as well as lost to the world, into her text. She takes him from obscurity by making him known. But more importantly, the narrator-writer Elaine

61 Quite literally, Mr. Potter is hardly allotted any direct speech throughout the novel. His utterances are mostly restricted to "eh, eh." Justin D. Edwards reads this as "an expressive lack – a lack of language, a lack of self-awareness, and a lack of understanding" (141). Julin Everett adds that "eh, eh" can be read as "an expression used in the Caribbean to indicate disapproval, pleasure, surprise, and curiosity, or to call attention to an object or to oneself. The lack of precision in Mr. Potter's utterances demonstrates the little importance he places on words, and thus, on defining his own identity" (59). More than that, Denise DeCaires Narain interprets the absence of Mr. Potter's voice as indicating a political problem: "Mr. Potter's 'insignificance' as a subaltern *man*, located on the 'fringes of the world,' is compounded by his willingness to accept this marginality with unquestioning equanimity. In response to his abject position, he appears to be capable only of making sounds, half-words and half-sentences which indicate no interest in, or understanding of his position" ("Writing 'Home'" 503, emphasis in original).

62 Braziel moreover reads *Mr. Potter* not only as the creation of a father figure, but more broadly as a Caribbean genesis myth: "*Mr. Potter* is a postcolonial, postmodernist creation myth – postmodern in its sensibilities (language constructs transitory truths), yet modern in its historical crises (the past haunts not only the fleeting moments of the present but also of the future)" ("Another line" 129), through which Kincaid "joins a Caribbean 'quarrel with history' that is above all a preoccupation with genesis (origins, creation, filiation)" ("Another line" 130).

Potter uses text to reverse power relations previously at the center of her relationship with her absentee father: he made her, but withheld himself, which put her at a disadvantage in that she felt the lack of filiation. Now, *she* creates *him*. Repetition bespeaks what hitherto was not, as is also the case in the daughter's fivefold narration of meeting Roderick Potter, as I will show in the following section.

Meeting Mr. Potter: Once, Never, Repeatedly

Kincaid's narrator in *Mr. Potter* not only calls her father into the text; through repetition she also constructs a memory of an encounter of father and daughter when she was a young girl. This meeting arguably might never have happened; or it might have happened but the memory thereof was distorted by temporal remove or by the mother's narrative interference. Nonetheless, Elaine Potter claims a memory of having met her father as her own. The repeated narration of the meeting turns Mr. Potter from an unknown and thus nonexistent entity into a father figure who the daughter interacted with – at least textually, when the event is narrated five times over the course of thirty-eight pages (*MP* 124–161). Each narration differs in detail and length. The daughter-narrator's writing and rewriting with a difference, I argue, establishes a connection with Roderick Potter, as well as it intensifies the familial relationship.

In her five narrations of (not) meeting Roderick Potter, the narrator-writer Elaine Potter "remembers" being sent to her father's place of employment, a garage at which he works as a chauffeur, in order to ask him for money for school supplies.[63] The first version of the episode is the longest of the five, extending over four pages (*MP* 124–127). In it, Elaine Potter establishes her encounter with Roderick Potter, while yet also focusing on the insurmountable space between them. She writes,

> when I was about four years old, I saw Mr. Potter standing in the space that was between the street and the entrance to the garage, and that was the first time I can remember seeing him standing between the street which led to me and the world beyond and the entrance to the garage, which held inside it all the darkness of the world when it has been reduced and made small and powered by evil. (*MP* 124–125)

63 Again, this memory is a mediated one, as Elaine Potter acknowledges: "I remember this incident of waving at him because my mother has told me about it and through my mother's words, I have come to see myself waving to Mr. Potter" (*MP* 127). Her mother's narrative involvement in Elaine Potter's memory is again mentioned twice on the following page, emphasizing the memory-making capacity of words, which the narrator equally makes use of here.

In Elaine Potter's perception and narration, spatial positions characterize the two parties involved: Mr. Potter is located in an interstitial space between the sunny street and the open garage gate which gapes like a black hole. The narrator's choice of words, referring to "all the darkness of the world," which is "powered by evil," here suggests the association of the garage with the colonial history of Antigua, which elsewhere is described as an evil force (*MP* 84). Behind the girl the street opens up to the rest of the world, from which Mr. Potter is excluded in his interstice between "the world" and "the darkness of the world" in the garage.

Perhaps inadvertently, young Elaine – or rather the adult narrator Elaine – attempts to bridge the gap between her body and her father's by claiming an affiliation: "I had waved at him, I had stood before him and wished him a good morning, and I had said, through gestures only, that he was mine and I was his, [...] that the seemingly invisible spaces between two people who shared a common intimate history were impossible to destroy" (*MP* 125). While this quote points to a hopefulness in the girl to build a connection with her father, it remains ambiguous in suggesting the very possibility of that. On the one hand, the "invisible spaces" that are "impossible to destroy" may be read as an indelible link. On the other hand, they may as well be understood as an impermeable barrier. In this, the spaces would not connect, but divide and keep apart. Ultimately, both of these suggested interpretations are equally viable, especially when thought together – what connects Elaine and Roderick Potter is their disconnection.

Mr. Potter's reaction to the girl's advance is one of utter disinterest. He does not even "dismiss [her] with a wave of his hand" (*MP* 126), but solely rolls his shoulders back and forth – a bodily movement with even less expressive power than the shrugging of one's shoulders. His daughter perceives his refusal to acknowledge her as worse than a dismissal: "Not only did he ignore me, he made sure that until the day he died, I did not exist at all" (*MP* 126). This nonexistence, however, is now challenged by her writing of the incident in which she connects her own person to his by writing about them both – albeit by describing the distance between them.

Twenty pages later, *Mr. Potter*'s narrator-author again writes the incident of meeting and at the same time not meeting her father. In the rewriting, the girl's encounter with him is even more unsuccessful than in her first version of the event. Instead of being positioned in the sunny street, she now is aligned with the darkness of shadows herself (*MP* 146) while waiting for Mr. Potter to ask him for "something essential, [...]. Schoolbooks, for instance" (*MP* 146). The girl then is informed by a co-worker of Mr. Potter's that he would not come, but accustomed to waiting for him already, she stays anyway. "[A]nd then Mr. Potter came, driving a car with the brand Hillman or Zephyr stamped on it, and when he saw me, he waved me away as if I were nothing to him at all and had suddenly and insanely decided to pursue an intimate relationship with him" (*MP* 146). Notably, the repetition of the incident takes its cues from the first, but with variations. The story stays the same – a girl tries to meet her biological father at his place of work in order to ask him for monetary support, but he disregards her. However, the differences in the repetition change the significance of the incident. In Elaine Potter's second narration, details become fuzzy. It is of little importance what exactly she is sent to ask for, or which car he drives. This is accompanied with changes in spatial order: a darkness encroaches from the father figure and the garage behind him towards the girl, which might symbolically denote an effect of the paternal rejection. Mr. Potter's response to the girl also intensifies in that he now actively rejects her. He communicates his rejection through gesture – one that acknowledges her no more than an abandoned dog, but also no less than that. In comparison with the first narration of the event Mr. Potter's rejection of his daughter intensifies. Elaine Potter's rewriting now includes a direct response to her claiming an "intimate relationship" with him, as she did in her first account of the meeting (*MP* 125). Elaine Potter's revisions also elicit a response from her father, such as gesturing towards her, which had previously been absent.

This development plays itself out further in the following three repetitions of the incident. The third, very short version again emphasizes the distance between father and daughter, when Elaine Potter recounts, "from across Redcliff Street, I, then a child with my name Elaine Cynthia Potter, saw Mr. Potter, [...] I had never met the man whose name I bore. I saw him from across the street and from across the street I asked him for money to buy books that I needed for school" (*MP* 153–154). Here, the space between the two is not crossed by communication or gestures. The repetition of "across the street" highlights the distance between father and daughter. No further communication takes place, which supports the daughter's assertion that she actually never met Mr. Potter.

The physical distance is then bridged by the fourth rewriting of the incident, which very shortly reads: "Mr. Potter slammed a door in my face when I was sent by my mother to ask him for a tablet of writing paper" (*MP* 157–158). This version is rather similar to the fifth and last in which the daughter-narrator declares: "I have only a vague memory of him ignoring me as I passed him in the street, of him slamming a door in my face when I was sent to ask him for money I needed to purchase my writing paper" (*MP* 160–161). With this last narration of the incident, it might be most obvious how much the five versions differ from each other. In the first one, any interaction between father and daughter is kept to a minimum with Elaine Potter only looking at Mr. Potter, and the father not even acknowledging the daughter. In the last version of the failed meeting, by contrast, his reaction is decidedly one of rejection. He ignores her just as she passes him, and although his reaction is not a favorable one, his ignoring her requires his recognition of her presence. More so, when she verbally approaches him, he reacts with force by shutting a door in her face and thus shutting her out.

Writing becomes the tool to turn the power relations between father and daughter, when she now claims the authority to write him. This is particularly taunting, as writing also was at the center of the failed encounter with Mr. Potter – he rejected her when she asked for money for paper. Within the very medium of writing the daughter now frames Mr. Potter as *she* pleases. In the course of the five narrative repetitions of the encounter, father and daughter do not overcome the distance between them, but the intensification of actions and reactions establishes a relationship between them that replaces the absence of any connection at all. Textually, Elaine Potter creates a memory of having made a connection with Roderick Potter, even if it is an unsuccessful one. Although he must remain absent and unknowable, in this last instance she has indeed met her father.

Dramatist Suzan-Lori Parks also makes use of repetition, both on word level and in the dramatic structures of her plays. In her essay "From Elements of Style," Parks elaborates her views on the technique, which she calls "Rep & Rev" and which is closely related to Gates's Signifyin(g): "Characters refigure their words and through a refiguration of language show us that they are experiencing their situation anew" (Parks 9). Through repetition, characters may then examine "something larger than one moment" (Parks 9).[64] Something similar is at play in Kincaid and Elaine Potter's refigurations of meeting Mr. Potter. In her narrative the daughter returns to the experience of being rejected by the father, which may be read as paradigmatic for the father-daughter relationship in *Mr. Potter*, as the text overall presents the daughter's simultaneous approach to and distancing from her absentee biological father.

In creating a memory of her father, Elaine Potter not only reverses the power relations between them, but she also emerges in the text herself. Just as she calls Mr. Potter into her text, Elaine Potter writers *herself* into being in her own work. Considering the five narrations of the girl (not) meeting Mr. Potter at the garage, it is noteworthy that while his reaction to her intensifies, also her presence in the text does.

In the first two narrations (*MP* 124–127 and *MP* 146), the focus lies on Mr. Potter and his position. Hers is foremost defined in relation to his. This changes in the third narration. To quote it again: "from across Redcliff Street, I, then a child with my name Elaine Cynthia Potter, saw Mr. Potter, […] I had never met the man whose

---

64 For insightful analyses of Parks's use of "Rep & Rev" to performatively investigate the history of slavery in her plays, see Ilka Saal's *Collusions of Fact and Fiction. Performing Slavery in the Works of Suzan-Lori Parks and Kara Walker.*

name I bore. I saw him from across the street and from across the street I asked him for money to buy books that I needed for school" (*MP* 153–154). In this version, the space that the girl occupies is the one identified by the text (as "Redcliff Street"), in opposition to Mr. Potter's garage in the first narration. Moreover, in the two previous versions, the narrator (and her younger self) were only identified by the personal pronoun "I." Now, as that "I" attains a defined spatial position, the pronoun is "filled," when the narrator ostentatiously mentions her own name "Elaine Cynthia Potter." Volunteering her full name, almost as if physically holding it in front of herself, the girl presents it to her father, just as the narrator-author presents it to her readers, and she claims it for herself. While this lays claim to Mr. Potter by the shared family name, the enunciation (and writing) of the name also defines the speaker of the text and establishes the girl as a subject.

At the same time, the use of the pronoun "I" increases. In the few lines quoted above, it occurs six times. Whereas elsewhere Mr. Potter's name echoes through the text, here it is distinctly the "I" that does. I argue that this both aligns the pronoun with the specific reference of the full name and establishes a space for "Elaine Potter" within the text. The repetition of "I" carves out a space from which the text speaks. So while the father Mr. Potter does not recognize Elaine Potter, the text at hand *does*.

The coalescence of name and role in Mr. Potter's case constitutes the whole of a specific signified, namely the thought construct of Mr. Potter as Elaine Potter's father. In the same way, the multitude of signifiers here ("I") transforms an impotent little girl into a narrator-author who inhabits the position from which the text speaks. The repetitions here constitute Elaine Potter as a subject with agency and the power to control the narrative.

This empowerment is also evident in Elaine Potter's rejection of her father's only inheritance: parentlessness. Like his daughter, Roderick Potter suffered the absence of his father. He was regarded as having "a line drawn through him, and by this it was meant that he had no father, no father's name was written in that column on his birth certificate, only a line had been drawn through it" (*MP* 97). The line is transferred from the paper to the person. Decoding this metaphor, one could say that Mr. Potter's whole existence is denied by the absence of his progenitor. This begs the question of how Roderick Potter came to acquire his father's family name in the first place. The text clarifies that "his name then was Roderick Nathaniel on his birth certificate, his name then was Roderick Potter in his mother's mind" (*MP* 79), which suggests that the family name was not assigned by law and by a registrar, but by his mother Elfrida Robinson who associated her son with his father.

Of her own birth certificate Elaine Potter writes that

> there is an empty space with a line drawn though it where the name of my father, Roderick Nathaniel Potter, ought to be […]. And this line that runs through Mr. Potter and that he then gave to me, I have not given it to anyone, […] I have made it stop with me, for I can read and I can now write and I now say, in writing, that this line drawn through the space where the name of the father ought to be has come to an end, and that from Mr. Potter to me, no one after that shall have a line drawn through the space where the name of the father ought to be, and that through him coming through me, everyone after that shall have a father and a mother. (*MP* 100–101)

The lines metaphorically running through Elaine and Roderick Potter stands contrary to a line of ancestry. Crossing the space on the official form, the lines cross out a possible line of names connecting past and present. The line thus presents a discontinuation of a family line. Yet the line itself is continued being ceded from parent to child through the absence of the parent. What indeed is inherited is a legacy of absence. To break with this legacy of nothingness, a presence is required, such as the one Elaine Potter vows to create by ensuring that her own descendants shall have both mother and father.

Moreover, the text that Elaine Potter produces in *Mr. Potter* discontinues the legacy of absence. The fictional account of Mr. Potter fills the empty space that the missing name of the father has left. In this, Elaine Potter's writing ties on to the line of ancestry she and Mr. Potter had previously lacked. By making her own father and thus reversing the order of genealogy, and by doing so in textual structures that deny the existence of a single continued (or discontinued) line of ancestry by exploring "possibilities of the real," Elaine Potter denies the legacy of absence and claims the name "Potter" as her own. In the absence of the father's name and recognition, Elaine Potter should likely have been named Elaine Richardson after her mother. Nevertheless, through text Elaine Potter assigns her own family name to herself and with that also her family affiliation. She thus filled the empty box where her father's name ought to be by rejecting the exact repetition of her father's family structures, both in writing and in her approach to her own descendants. Rather, she opts for a different outcome and creates it herself.

Ultimately, in these examples from *Mr. Potter* repetitions with a difference work out new meanings to the signs. The back and forth – or "hall of mirrors," as Gates phrases it – posits meanings that hitherto were beyond the surface, or even nonexistent. This technique of redirection gives the narrative a sense of impermanence. The continuous movement of the text that opens up spaces for creation, such as that of a father figure on the terms of a daughter who can claim the power to write him and herself into being and who can envision a different future for herself and her own descendants.

2.2 Writing with and through Other Texts: Intertextual Engagements with History

Jamaica Kincaid's oeuvre is fundamentally characterized by elaborate forms of referencing that work out new positions on seemingly established knowledge. This chapter on intertextual relations provides insights into the workings of textual referencing of other or earlier texts and how Kincaid's works thus engage both literary history and historical discourses.

As Renate Lachmann points out,

> the original idea of making literature means first of all making literature *from* literature, that is, writing as continuation, writing as rejoinder, or rewriting. This notion of making literature out of literature has important implications for both the discreteness of any individual text and for its closure and 'totality.' Instead of a process of selection and combination that presupposes a grammar of forms, we are now confronted with a ramifying growth of meaning that proceeds from the interfacing of texts. (37, emphasis in original)

According to Lachmann's understanding of text, every text is always already entangled with other texts. This closely resembles Roland Barthes' conception of text as a web of structures, meanings, and other texts. Kincaid's texts pick up on these notions overtly, particularly in the latest novel *See Now Then*, which explores the simultaneous making of text and past through the metaphor of knitting (see chapter 1). In what follows, this chapter now explores the "woven" or "knitted" nature of Kincaid's other texts in more detail, particularly concerning the interplay of intertextuality, referentiality, and history.

Lachmann argues that because every text entails the heritage of other texts, "[t]he memory of a text is its intertextuality" (15). In my inquiry into textual engagements with history, I ask how this memorial aspect of intertextuality occurs in Jamaica Kincaid's texts. Rather than identifying individual intertexts to Kincaid's work – a task which has by now often enough been accomplished by other scholars[65] – I investigate how other texts and discourses are taken up in Kincaid's contemporary texts and what this play with reference discloses about the texts' conception of history.

The first section of this chapter, "Locating Kincaid's Works in a Web of Con-Texts," maps the literary and cultural contexts of Kincaid's work between the Caribbean and the United States. The following section is concerned with historical colonial discourses which Kincaid's novels *Mr. Potter* and *See Now Then* take up via

65 See, for instance, Chansky, Gass, Pelt, Rodriguez, or Simmons ("Jamaica Kincaid and the Canon").

intertextual engagements with earlier (literary) texts. Both novels present how history and historiography affect contemporary thought and writing, which in *Mr. Potter*'s case is exposed through textual mimicry, and which is revised through resignification in *See Now Then*. The third section of this chapter on "Heracles and Persephone" then investigates *See Now Then*'s engagement with Greek mythology and how intertextuality here facilitates the integration of contemporary literary and social structures that are based on colonial history into a web of larger historical contexts.

Locating Kincaid's Works in a Web of Con-Texts

Since texts should be read in their con-texts,[66] as Lachmann's conception of the making of text and Barthes's textile metaphor of texts as web-like structures suggest, this section considers the general contexts of Kincaid's works – in terms of the cultural and political contexts of their emergence as well as their places in Caribbean and North American literary histories – to unpack the breadth of the textual-cultural fabric into which, through which, and from which Kincaid's texts are woven.

Jamaica Kincaid was born on May 25, 1949, in St. John's, Antigua, as Elaine Potter Richardson. She grew up on the island, whose colonial history is already signaled by the name – Santa Maria la Antigua – given to it by one of the most famous Western colonizers, Christopher Columbus, upon passing the island at a distance. Since English colonists settled the island in 1632, it remained under British control. Tobacco and sugarcane were the main crops produced on plantations, for which enslaved Africans were imported. Slavery was abolished in the British colony in 1834. After becoming a self-governing state of the United Kingdom, in 1981 the country of Antigua and Barbuda gained independence from British rule but still remains part of the Commonwealth of Nations.

66 The hyphenated spelling of "con-texts" here refers to John Thieme's variant spelling which seeks to "indicate postcolonial texts that engage in direct, if ambivalent, dialogue with the canon by virtue of responding to a classic English text" (4). Moreover, the syllabic separation highlights the prefix "con-" and hence the Latin origin of the word, which stems from "com" – with, together, and "texere" – to weave, make. I emphasize the con-/com- in "context" to highlight the coexistence of texts and discourses, be it contemporary ones or of earlier or later origin, in contrast to a notion of intertextuality that assumes the derivation of later texts from "parent texts," indebted to their predecessors for their own genesis and meaning-making properties.

Elaine Richardson's childhood and youth were spent in a British colony, well before its independence. As a girl, she received formal school education from the age of three and a half years until she was fifteen.[67] "In my generation," Kincaid said later, "the height of being a civilized person was to be English and to love English things and eat like English people. We couldn't really look like them, but we could approximate being an English person" (in Garis). This was also reflected by the curriculum of "Princess Margaret School," which Elaine Richardson attended: British national songs, such as "God Save Our King" or "Rule, Brittania" were as much part of education in Antiguan government schools as was canonical literature by authors such as William Shakespeare, John Milton, the Brontë Sisters, Thomas Hardy, or John Keats (see Kincaid in Cudjoe 397–398). Kincaid says about her school education that it "was very 'Empire,' [it] only involved civilization up to the British Empire – which would include writing – so I never read anything past Kipling. Kipling wasn't even considered a serious writer" (Kincaid in Cudjoe 397). Kincaid claims to never have read nor even known of works by Caribbean writers before leaving Antigua[68] and regards her early readings of the British literary canon as formative for her own literary endeavors.[69]

67 Elaine Richardson began to attend school at this early age under the pretense of already being five years old as her mother wanted her out of the house during the day, as she claims (Kincaid in Cudjoe 397).

68 In a 1994 interview with Moira Ferguson, Kincaid relates: "I had never read a West Indian writer when I started to write. Never. I didn't even know there was such a thing, until I met Derek Walcott. He said 'Do you know – ?' […] I did not know. So I do not come from the West Indian writing tradition, and there is no such thing. There might be, eventually, but we are of the English-speaking tradition" ("A Lot of Memory" 169). For instance, scholar Carole Boyce Davies regards these claims of Kincaid's on Europeanness instead of on a Caribbeanness as highly problematic (expressed for instance at the conference "The Art and Craft of Grafting in Jamaica Kincaid's Works" in Paris, 2017). Indeed, it is somewhat baffling that Kincaid, who in other aspects insists on her individuality instead of being grouped with any kind of genre or political orientation, here self-identifies with a writerly tradition. Henry Louis Gates, Jr., in contrast, maintains that "Black writers, like critics of black literature, learn to write by reading literature, especially the canonical texts of the Western tradition. Consequently, black texts resemble other, Western texts and employ many of the conventions of literacy form that comprise the Western tradition. Black literature shares much with, far more than if differs from, the Western textual tradition" (*Signifying Monkey* xxii). Moreover, Kincaid's insistence on the effect of colonial education also critiques colonial assimilation and highlights the opportunity of reversing the colonizing gaze/writing tools, as Kincaid does in her works.

69 Kincaid's works are just one example of (post)colonial writers being affected by the British canonical texts. The same references occur in the biographies of other postcolonial authors: for instance, just as Kincaid regards British writer of children's books Enid Blyton as her first role model (Kincaid in Galassi), so does Nigerian writer Chimamanda Ngozi Adichie, who recalls: "When I started to write, I was writing Enid Blyton stories" (qtd. in Boyagoda). And just as Kincaid's

While heavily affected by British colonial education, Elaine Richardson also grew up in a space determined by the Caribbean spiritual practice of obeah, a belief rooted in African traditions.[70] Obeah "is a system of beliefs rooted in Creole notions of spirituality, which acknowledges the existence and power of the supernatural world" (Fernández Olmos and Paravisini-Gebert 131). Both Richardson's mother and grandmother were believers and practitioners of obeah (Simmons, *Jamaica Kincaid* 8). Kincaid later described obeah's approach to the world as "chaotic" (Simmons, *Jamaica Kincaid* 8) and when asked about elements of magic realism in her work, she explained: "*The truth is I come from a place that's very unreal.* [...] The place I come from goes off in fantasy all the time so that every event is continually a spectacle [...]. I'm not really a very imaginative writer, but the reality of my background is fantastic" (in Vorda 13, emphasis in original). A certain degree of being comfortable with the unreal, or with a coexisting variety of realities is also noticeable in several other interviews with Kincaid in which she claims that "truth often makes for a better art. [...] [T]he truth, incidentally, is multifaceted. A lie is one thing, one single thing" (Kincaid in Buckner 465–469).[71] In Kincaid's works, this notion of "the real" is evident in her narrative repetition which continually provides different perspectives on past events (as the previous chapter on *Mr. Potter* has shown), or in Lucy's search for a hidden reality not visible to the naked eye, which she seeks to uncover when developing her photographs (see chapter 3.1). Modernist works approach the multifaceted representation of assumed reality in a similar fashion. Kincaid mentions writers such as Virginia Woolf, James Joyce, Alain Robbe-Grillet, or Nadine Gordimer as inspiring examples: "When I read them [here specifically Robbe-Grillet's short stories], the top of my head came off and I thought, 'This is really living!' And I knew that whatever I did, I would not be interested in realism" (Kincaid in Cudjoe 403). While realism claims faithfulness in representation, to Kincaid it is modernist writing that textually approximates her experience of life. Hence the Caribbean cultural practice of obeah with its fluent boundaries between reality and fiction, and writing styles of modernism

---

characters Annie John and Lucy wish to change their given names to "Enid," "after the authoress Enid Blyton" (*Lucy* 149), "the author of the first books I [Annie] had discovered on my own" (*AJ* 51), so is the same-aged female protagonist of another postcolonial novel deeply affected by Enid Blyton and the Brontë sisters: Tambudzai in Zimbabwean Tsitsi Dangarembga's *Nervous Conditions* (93). Kincaid, too, claims Charlotte Brontë as her second early role model (in Galassi).

70 Different scholars attribute the origins of the term "obeah" and of the practice are attributed to either Akan, Twi, or Igbo. See Rucker 40.

71 Kincaid makes similar statements on truth in other interviews, for instance with Perry, Sela, and Bonetti.

which are widely regarded as coming from a European tradition, converge in the rejection of singular truths by both Kincaid and her works.

Elaine Richardson left Antigua in 1965 for New York City at the age of sixteen to follow a common trajectory of immigrant workers in seeking employment as a nanny for an affluent white family in order to support her family in her native country with remittances. After her arrival in the United States, Elaine Richardson went to school at night and eventually obtained a high school diploma. Rather than continuing on the trajectory her family had projected for her, Richardson broke off contact with them; and instead of attending nursing school, she began to study photography in New Hampshire. Her early interest in visuality and particularly in photography continues to play a key role in Kincaid's works, almost all of which interrogate ways of seeing by referencing or even including photography.

A year later, Richardson returned to New York in search of opportunities to earn money through writing.[72] She became acquainted with the New York literary scene of the early 1970s, which eventually led to her writing for *The New Yorker* in 1974, where she obtained a staff position in 1976 that she held until 1995.[73] Concurrently with the onset of her professional writing, in 1973 Elaine Richardson changed her name to Jamaica Kincaid in order to both claim her Caribbean heritage as well as to sever her ties from it. In an interview with Allan Vorda, Kincaid explained that she took on a new name to conceal her literary endeavors from the people she knew. On the one hand, she sought the freedom to take a shot at writing without possibly being ridiculed in the case of failure. On the other hand, Kincaid also desired anonymity to freely express her thoughts about both Antigua and herself: "I wanted to speak truthfully about myself without being myself" (in Vorda 15). While the name change thus foremost appears as a disguise and hence a device to distance herself from her origins, the choice of name also reflects a claim on precisely these origins, as Kincaid says that "[i]t was a kind of invention: I wouldn't go home to visit that part of the world, so I decided to recreate it. 'Jamaica' was symbolic of that place" (in Cudjoe 400). Elsewhere Kincaid appears to delight in the etymology of her chosen first name when she explains that "Jamaica" is an English corruption of "Xaymaca" (in Vorda 15), the original Arawak name of the island. In contrast to a majority of the Caribbean islands, Jamaica kept its Indigenous name (though Europeanized). Kincaid may thus delight both in the persistence of Arawak Indigeneity and in the reflection of colonial history in the corruption of the island's name. The choice of her new last name Kincaid nonchalantly

72 Her very first published piece was an interview with Gloria Steinem for the magazine *Ingenue* in 1973.

73 Kincaid's very early work for *The New Yorker* mostly consisted of unsigned pieces for the column "Talk of the Town," which were anthologized in *Talk Stories* in 2001.

presents as rather arbitrary, when she asserts that it just seemed to go with "Jamaica" (in Vorda 15 and in Cudjoe 400). The change from Elaine Potter Richardson to Jamaica Kincaid highlights a desire for self-making as well as a double-bind conspicuous throughout the author's works that oscillates between approximation and distancing of the familiar and the unfamiliar, of the loved and the detested, of the colonized and the colonizing, all of which often converge in one.

In 1983, Kincaid's first book *At the Bottom of the River*, a collection of short stories, was published to great critical acclaim. This was followed by the author's first novel *Annie John* in 1985 and the nonfictional *A Small Place*, which resulted from her first visit to Antigua in the mid-1980s after nineteen years of absence from the island. These works all were written and published in the United States, and after Kincaid had lived there for almost two decades. Writing in the United States, but foremost concerned with the Caribbean in her texts, Kincaid's migratory position is not uncommon to Caribbean writers of her generation.[74] Frank Birbalsingh thus counts Kincaid among a younger generation of Caribbean writers who became known in or after the 1970s.[75] Critical for Birbalsingh, however, is not the decade

74 Compare, for instance, Olive Senior, Dionne Brand, or Michelle Cliff.

75 Tobias Döring postulates a generational divide between Caribbean writers with a similar timeframe, although he differentiates the generations based on other premises: in his study *Caribbean-English Passages: Intertextuality in a Postcolonial Tradition*, Döring is concerned with the cross-fertilization of canonical British texts and Anglophone Caribbean literary works. He regards, for instance, Derek Walcott (b. 1930), Wilson Harris (b. 1921), and V. S. Naipaul (b. 1932) as the "pioneering father figures in Caribbean literature" (10), who had to grapple with "the profusion of previous writing," by which Döring refers to the overbearance of English literary traditions and models (10). A younger generation of Caribbean writers, among which Döring counts for instance David Dabydeen (b. 1955), would have accordingly worked "in an already established Caribbean-English writing tradition" (Döring 10). Döring explicitly excludes female authors from this generational formation of Caribbean writers. He justifies this with reference to M. NourbeSe Philip, whom he reads as claiming that "for female Caribbean writers like herself the premise of literary production lies in 'the absence of writing'" (Döring 10 and Philip 76). However, while Philip laments the absence of an Afro-Caribbean writing tradition (76), similar to what Döring reads in the trajectories of the "father figures of Caribbean literature," Philip describes the struggle with the English language, which she regards unfit to express the fullness of the African-Caribbean experience (81–82). "That silence has had a profound effect upon the English-speaking African Caribbean artist working in the medium of words," Philip diagnoses (82). "The Absence of Writing" to which Philip points with the title of her essay is thus caused by the overbearance of the English language. Moreover, a consideration of Kincaid and other female postcolonial authors mentioned disproves Döring's assertion of the absence of literary ancestry for women writers: even though Kincaid testifies to an absence of Caribbean writerly role models, comparable to what Döring claims for Walcott et al., she does explicitly reference British canonical works, such as those by the Brontë sisters, Milton, or Blyton as texts by which she was thoroughly affected as a young reader and writer. This shows that pace Döring, both male and female Caribbean authors had to and have to navigate both the

of the 1970s itself, but independence from colonial rule which was achieved by most Caribbean countries in the 1960s. According to Birbalsingh, this induced a shift in writerly perspective: while Birbalsingh sees the previous generation of postwar and preindependence writers, such as Edgar Mittelholzer, Sam Selvon, George Lamming, or V. S. Naipaul as concerned with the ills of colonialism, he notes that younger Caribbean writers would focus on the postindependence experience. During her interview with Birbalsingh, Kincaid confirms that unlike for instance George Lamming, she does not rebel against "the great United States," but "it is about the things that the powerful United States can do and does do" (142). The evolution in Caribbean author generations that Birbalsingh discerns thus reflects a move away from the rebellion against monolithic colonial power structures towards more differentiated critiques of social and political issues of postindependence Caribbean societies.

Moreover, Birbalsingh diagnoses a fourth generation of writers, which he calls "diaspora of Fourth Stage writers" (142), among which he also counts Kincaid. While the author herself rejects the label at the time of the interview in 1991,[76] as she did not consider herself "American" enough to be considered diasporic (in Birbalsingh 143), her work indeed is affected by her move to the United States.

With this varied field of literature in which Kincaid locates her own writing, from colonial British education and canonical literary works to obeah to modernist literature to the study of photography and the general circumstances of life in Antigua as well as in the United States, it is befitting that Kincaid herself "claim[s] the right to ambiguity, and the right to clarity. […] I feel free to use everything, or not, as I choose" (Kincaid in Bonetti 129). It is hence unsurprising that Jamaica Kincaid is regarded both as a Caribbean and as an American writer, as for instance evidenced by the inclusion of her biography and works in standard anthologies or companions such as the *Encyclopedia of Caribbean Literature* and *The Routledge*

---

excess and the lack of literary models. Born in 1949, Kincaid would fall into both generational categories Döring postulates. Yet while Kincaid's age might place her among the younger generation of writers and earlier male and/or female Caribbean authors' texts already existed in her time, she claims she was not aware of them because she had had no access to their writings. Regarding the processes of becoming a writer, she would thus have to be counted among Döring's older and pioneering generation of Caribbean writers, who had to grapple with English tradition, while she was also becoming a writer herself.

76 This would change a year later. In 1993, Kincaid became an American citizen (Dance 152) and later in life she revealed how she was intrigued by African American culture and life (Rostron). Moreover, she converted to Judaism (also in 1993). She says: "I was trying to get the children to integrate all the different strands of their ancestral memory – that my family came from one part of the world through a certain set of historical events; their father's [Jewish] family, through a certain set of events came from another part of the world and had its own ancestral memory" (Sela).

Reader in Caribbean Literature, or *The Norton Anthology of American Literature* and *The Oxford Encyclopedia of American Literature* and *The Oxford Companion to Women's Writing in the United States*, or *The Cambridge Companion to the African American Novel* and *The Norton Anthology of African American Literature*.[77]

Ultimately, both Kincaid's biography and her literary texts indicate that Kincaid's works should not be read in single (and singular) contexts, but the multiplicity of cultural influences on the person and on the texts introduce multitudes of interpretational approaches to her literary works. Some of the influences which occur in the form of intertextual engagements with other texts and discourses in Kincaid's works are at the center of the following sections, which offer analyses of intertextual references particularly in Kincaid's latest novels *Mr. Potter* and *See Now Then*.[78]

Colonial Discourses: Visiting and Revising Textual Sites of History

In their engagements with history and historiography, Kincaid's texts take up colonial discourses to expose them as extending to contemporary textual production. I examine this in two exemplary readings of *Mr. Potter* and *See Now Then*. *Mr. Potter* is concerned with Antiguan colonial history and highlights the continuing effects of colonial sentiments on present-day conceptions of Antigua and its inhabitants by mimicking such discourses. *See Now Then* takes this a step further by reversing a literary motif of the nineteenth century and thus presents and questions seemingly established structures in racial relations. Both texts initially appear to repeat colonial texts rather uncritically. However, in *Mr. Potter*, the mimicking of intertexts exposes the totalizing claims of Western historiography, which is countered by the narrative of *Mr. Potter* itself. Similarly, in *See Now Then*, European claims to interpretative power over the bodies of others and of their textual

77 These examples were chosen at random and do not reflect the full spectrum of anthologies including articles on Kincaid and her works.

78 See also chapter 3.1 and chapter 3.2 for discussions of intertextual referencing of historical events of Columbus's travels in *Annie John*, and garden and travel writing in *My Garden Book:*, respectively. Generally, the fact that Kincaid's works are replete with references to other texts has already been demonstrated by the works of several other scholars. In what follows, I will therefore not attempt to provide a comprehensive overview of other (literary) works that found entrance to Kincaid's texts, but rather expose the functions of intertextual referencing in two choice novels by the author in which the referrals to earlier texts specifically amount to engagements with (literary) history.

representations are deconstructed when other perspectives reveal different histories.

In *Mr. Potter* the eponymous main character is displayed as affected by colonial history. Equally, the textual conception of this African-Antiguan character is written through colonial discourses. Intertextual references to Georg Friedrich Wilhelm Hegel's assessment of African history and people and by extension also their descendants, such as Antiguan Mr. Potter, heavily bear on the descriptions of the novel's protagonist.

While actual events in the history of transatlantic slavery destroyed actual people in removing them from their homes and turning them into chattel, people were moreover obliterated by History with a capital *H* – their own histories were hardly preserved, nor was their humanity recognized in the writing of Western history that still continues to obliterate people, as is exemplified in the literary figure of Mr. Potter. What is presented as the effect of the past on the character Mr. Potter, ultimately is the effect of text on text in *Mr. Potter*.

Perhaps the most prominent feature of the novel's characters is their utter unimportance. In an apparent paradox, the text centers on people it presents as hardly worthy of narration. For instance, the narrator-author writes of Elfrida Robinson, mother to the novel's protagonist Mr. Roderick Potter: "[I]t [Antigua, home to Elfrida] was only the land of a very small island, an island of *no account*, really, and she was of *no account*, really, only she was the mother of my father and I know I cannot make myself forget that" (*MP* 73, emphasis mine). The reiteration of being of "no account" here points not to the narrator's own evaluation, but to an absence of importance conferred by textual treatment. The "account" here doubles in its meanings of "importance" and "narrative." Elfrida – and with her, any descendants of enslaved Africans (*MP* 85), or generally the inhabitants of Antigua (*MP* 121) – are described as not important enough to have accounts, i. e., narratives, of themselves. With this, the narrator Elaine Potter indicates the nonhistory of the Caribbean as it manifests with regard to personal histories. Nonetheless, she now writes of them, and she does so to sustain her own personal history. The quote above thus presents the central concern of *Mr. Potter*'s narrator: in her emphasis on absence, she bespeaks it and thus dissipates it (at least to some extent) for her very own personal interests in history.

Ostensibly, *Mr. Potter* generally is concerned with "nothing." The word is uncannily present throughout the text – people are "nothing" to each other (*MP* 150); Mr. Potter thinks of "nothing" (*MP* 27 or 34); fish pots are "filled with nothing"

(*MP* 43); songs sung mean "nothing" (*MP* 140).[79] The narrator's observations of the ever-present "nothing" commonly refer to everyday Antiguan life. Harsher uses of the term are applied by Dr. Weizenger, a newly arrived immigrant who had fled from the European holocaust against the Jewish people. Although just having survived the attempted eradication of a whole people, Dr. Weizenger does not see Mr. Potter as an equal survivor of history:

> Mr. Potter, the entity that made up Mr. Potter, was nothing itself, nothing in the sense of something without worth, nothing in the way a lighted matchstick when it is not needed, so Dr. Weizenger thought and so too thought the rest of the world, the rest of the world who could have an idea in regard to anything and then launch that idea into the realm of the everyday. (*MP* 19)

The European Dr. Weizenger is aligned with those who have the voice and agency to speak and to make their opinions known. In that, he represents a generalized Western opinion of people like Mr. Potter.[80]

John Clement Ball contextualizes this overly present addressing of "nothing" with the absence that echoes through historiography: "With the constant repetition of 'nothing,' Kincaid echoes an enduring rhetoric of negation in Caribbean discursive history that dates back to writings of the earliest European visitors" (206). With "rhetoric of negation," Ball refers to the repeated abjection of Indigenous peoples, enslaved Africans, and English colonizers who settled in the Caribbean, in written accounts of this area, which deny the inhabitants of the Caribbean intelligence, morality, civility, and nobility. Ball illustrates this "rhetoric" with examples from travelogues and histories by the Dominican Tomás Ortiz (early sixteenth century), Edward Long (1774), James A. Froude (1888), and V. S. Naipaul's dictum that "nothing was created in the West Indies" (Ball 206–207). I add here Hegel's notorious take on the history of African peoples, which operates with the same rhetoric. Africa, according to Hegel, "is no historical part of the world; it has no movement or development to exhibit. [...] What we properly understand by Africa,

79 These are just a few examples of the occurrences of "nothing" in the text; the word features every few pages throughout the novel.

80 We might read this as suggesting a disparity in the historiographic treatment of the Jewish holocaust in comparison to the transatlantic slave trade. Although both Africans and Jews suffered at the hands of racist Western Europeans, the history of African enslaved people here remains "nothing." Although displaced, Dr. Weizenger retains the ability to speak, while Mr. Potter must remain silent. Curdella Forbes notes that Mr. Potter literally does not attain a voice here: "Mr. Potter's agency as the subject of all the narrative sentences is erased by the fact that he speaks fewer than ten sentences, all of them rendered in reported speech or in translation from his Creole utterance, the Creole itself displaced by appearing in parentheses after the translation" ("Fracturing Subjectivities" 32).

is the unhistorical, undeveloped spirit, still caught in the conditions of mere nature" (91–92), and he sees its inhabitants as "capable of no development or culture, and as we see them at this day, such have they always been" (91). In denying African history, Hegel attributes the continent with a different temporality altogether, one that passes slower and thus is marked by backwardness. The assumed absence of history transfers to African people in an assumption of inability of development, in Hegel's view. Thinking this together with the rhetoric of negation, which Ball attests to *Mr. Potter*, it points to a consistent and unrelenting monotony of life in the Caribbean.

This is also taken up in Elaine Potter's comments on "ordinariness" and "indifference" in addition to the prevailing nothingness that pervades Mr. Potter's life since birth. In 1922, Roderick Potter was born into a world that does not take any interest in this event: "And Mr. Potter was born, and all the world was indifferent to this" (*MP* 66). His mother Elfrida Robinson did not care, and neither did the midwife. Even the water the newborn was bathed in "was indifferent as to temperature" (*MP* 67). Generally, Mr. Potter's world is characterized by ordinariness and indifference. For instance, to his father, Nathaniel Potter, each day is ordinary (*MP* 58); the grave master who later shows the narrator-daughter to Mr. Potter's (possible) burial site is likewise an "ordinary man" (*MP* 49); all of Mr. Potter's daughters carry "ordinary names just like ordinary people" (*MP* 120); and the descendants of enslaved Africans are described as "of the ordinarily degraded" (*MP* 80). Events such as births or deaths do not make a difference here. When Mr. Potter's mother commits suicide by drowning herself, "the seas took her in, not with love, not with indifference, not with meaning of any kind" (*MP* 76). Elfrida is rendered as so unnoteworthy that she is not even met with indifference.

The emphasis on the ordinary in *Mr. Potter* paints a picture contrary to the development and movement Hegel regards as prerequisite for historicity. In this, narrator Elaine Potter repeats the ascription of an absence of eventfulness, which constitutes the absence of history in Hegel's view. In her repetition of this rhetoric of negation, Kincaid's text produces an excess of nothingness. "Nothing" is addressed to such an extent that this constitutes a presence that opposes the absence proclaimed by Hegel and others. The mimicking of colonial rhetoric simultaneously reverses it, as its continual references to nothingness transform the negation into a demonstration of colonial historiographic processes. In this way, Kincaid's text exposes that unimportance is indeed *not* an inherent characteristic of African people, but that processes of Western historiographic negation marginalize Black people. Homi Bhabha theorizes such repetition that already contains the subversion of the imitated as "colonial mimicry." On the side of the colonizer "colonial mimicry is the desire for a reformed, recognizable Other, *as a subject of difference that is almost the same, but not quite.* Which is to say, that the discourse of mimicry is constructed around an *ambivalence*; in order to be effective, mimicry must continually produce its slippage, its excess, its difference" (122, emphasis in original). As such, the repetition with a difference that is at the heart of colonial mimicry, always already undermines the mimicked. "The effect of mimicry on the authority of colonial discourse is profound and disturbing" (Bhabha 123). Bhabha calls this the "menace of mimicry," by which he refers to "the disclosing [of] the ambivalence of colonial discourse [which] disrupts its authority" (126). This is also at play in *Mr. Potter*'s rhetoric of negation, which through repetition positions absence front and center in the text, calling attention to the historic negation of people in the Caribbean and of African descent.

The mechanisms of the menace of mimicry are also apparent in Elfrida Robinson's naming of her son: overwhelmed by the "great expanse of the life of George Brydges Rodney, the English admiral" (*MP* 64), Elfrida is inspired by his name. The narrator of *Mr. Potter* deems this British naval officer known for plundering, and who was stationed in the Caribbean among other locations throughout the British Empire, a "maritime criminal, [...] whose criminal nature and accomplishments had become so distorted in retelling that the victims of his actions had come to revere him" (*MP* 64). Elfrida's admiration for a character such as Rodney once more highlights the effect of colonial ideology and historiography on the colonized. For her son, however, "whose appearance in the world had no real meaning for her, she wanted a name that had no meaning at all to her, and this wanting of no meaning made her choose something different, and so she called him Roderick, not Rodney" (*MP* 64). In naming her son after a British hero, but not quite as she wishes to remain in the realm of meaninglessness, Elfrida performs an idealized form of colonial mimicry that attempts to approximate Englishness, but has internalized a presumed inferiority. Ironically, Elfrida's choice then subverts the assumption of inferiority, considering that Roderick, a Germanic name, actually means "famous ruler." As such, the mimicking of the name with a difference actually trumps it, when she does not name her son *after* the famous ruler Rodney, but she proclaims him a famous ruler *himself*.

This ordinariness and Elfrida's intended absence of meaning recalls Hegel's negation of African history, which he extends to an assumed inferiority of African people. Hegel maintains:

> In Negro life, the characteristic point is the fact that consciousness has not yet attained to the contemplation of any firm objectivity – as for example, God or law – in which the interest of man's volition is involved and in which he has the view of his own being. In the indiscriminate, compact oneness of his existence, the African has not yet attained the distinction between himself as an individual and his essential generality [...]. The Negro as already observed, exhibits the natural man in his completely wild and untamed state: one must abstract from all thought of reverence and ethics – all that we call feeling – if we would right-

> ly comprehend him; there is nothing harmonious with humanity to be found in this type of character. (86)

Hegel's racist opinion here relegates Black people to an animal-like state of backwardness. Hegel locates an explanation for general shallowness of character in his assumption of an underdevelopment of African people, which links typicality (in opposition to self-conscious individuals) to the absence of European conceptions of history.

Mr. Potter picks up on this assumed absence of consciousness in its presentation of its protagonist as lacking the ability of self-reflection: "Mr. Potter did not think that any part of him was reflected in [the] sight before him [...] none of this [what he saw] reminded him of himself in any way and that was only because everything he saw was so closely bound up to him and all that he saw there was no distance of any kind" (*MP* 5).[81] In a Lacanian reading, Mr. Potter would have skipped the mirror stage, as he does not recognize the relationship between himself and his surroundings ("no distance") and for that matter he does not *reflect* on himself either: "he did not think [...] at all, for he was not at that moment separated from himself" (*MP* 21). According to Lacan, the mirror stage is imperative in the formation of the I, i.e., in attaining one's own subjectivity, which is thus denied Mr. Potter and others in the same situation. Roderick and Nathaniel Potter in this way just *are*; they are one with themselves (see *MP* 24, 45).[82] Curdella Forbes assesses that "Mr. Potter is ironically, and contemptibly, at one with the natural universe, not because he has attained a transcendental understanding of his relation to it, but because he has failed to attain the basic condition for humanity – the capacity for self-reflection. Self-reflection recognizes one's relation to the cosmos and a radical reflection from it; it is the arrival of is-ness, of subjectivity" ("Fracturing Subjectivities" 32–33). This "oneness" that results from the absence of reflection seems to directly take up Hegel's assumption of an African "compact oneness of [...] existence" (86). In the general climate of ordinariness and indifference, Mr. Potter cannot emerge as an individual character in his daughter's narration, but he re-

81 Similarly, Roderick Potter's father, Nathaniel, does not differentiate between himself and the world he is a part of, when he "looked outside himself and he looked within himself [...] it was all the same" (*MP* 45).

82 The rhetoric of negation also participates in the frequent descriptions of Mr. Potter's illiteracy, which once more emphasizes his unreflectiveness: "[B]ecause Mr. Potter could neither read nor write, he could not understand himself, he could not make himself known to others, he did not know himself, not that such things would have brought him any amount of happiness" (*MP* 21). The equation of illiteracy and unreflectiveness moreover recalls colonialist notions rendering cultures of orality subordinate to literate Western cultures.

mains a "type," being of no account and unable to give an account of himself.[83] His daughter's textual exploration of the past provides an explanatory model of his absence from his own life.[84] When her characterization of Roderick Potter takes up colonial discourses on Blackness, she shows how they still determine the historiographic writing of the Caribbean and of diasporic Black people. In imagining her father through Hegel and the like, the narrator draws attention to the totality of Western history in Glissant's sense, which excludes nuanced accounts of the descendants of enslaved people. In this way, she criticizes not only slavery and colonialism as they existed, but also their treatment in historiography, which still corrupts contemporary knowledge and imaginations of the past.

While the narrator of *Mr. Potter* repeats the "rhetoric of negation" of the Caribbean, she simultaneously denounces it in confronting it with her own voice. By imagining her father's life, the narrator-author creates knowledge about the past that opposes the previous nothingness and silence. Her mimicking of colonial prejudices towards African people, and towards the utilization of misguided opinions like those that link illiteracy, consciousness, and humanity, frames her own narrative within a critique of colonialism and slavery, demonstrating its aftereffects as the causes for conditions that prevail today. Her repetition of colonial negation of Antiguans thus entails a "double articulation" (Bhabha 122), which exposes and disturbs colonialist literary and historiographic dominance. Within her intertextual acts of mimicry, the narrator of *Mr. Potter* carves out space for herself in writing a version of history, which narrowly resolves the colonial negation of her ancestors: through her writing, the narrator-author presents her own agency to break her ancestors' cycle of silence, as she is able to give an account of Mr. Potter's life story. Though the narrator does not give a voice to the silenced Mr. Potter, she gives one to herself in producing *Mr. Potter.*

In a similar way, *See Now Then* voices matters that in colonial textual production remained mute. This novel makes use of intertextual references in order to evoke the textual history of the issues it is concerned with. However, unlike *Mr. Potter, See Now Then* not only exposes and challenges its textual ancestry by mim-

---

83 This is exemplified both in Roderick Potter and in his father Nathaniel: "And if someone came upon him then, he could not give an account of himself, not even one that began with 'I was born…' for he no longer had an interest in when he came into the world" (Nathaniel Potter, *MP* 42). Roderick Potter: "[H]e did not have a mind's eye in which he could wander, he had no thoughts about his past, his future, and his present which lay between them both […] he could not read and could not write and he could not render the story of life, his own in particular" (*MP* 130).

84 Julin Everett also reads *Mr. Potter* as an autobiographical orphan narrative in which the narrator creates her father in writing to thereby become herself, i.e., to overcome the void of that missing personal history.

icking it; the later novel also reverses colonial meanings by literally making it possible to read what has hitherto been opaque. In what follows, I show how *See Now Then* visits and revises two earlier novels, *Wide Sargasso Sea* by Jean Rhys (1966) and *Jane Eyre* by Charlotte Brontë (1847). These two texts are already intertextually connected, as Rhys conceived her novel as a prequel to Brontë's nineteenth-century text in order to remove the character of Bertha Mason, who is famously known as the "madwoman in the attic,"[85] from this position and from obscurity by giving her her own life story.[86] *See Now Then*, I argue, enters a conversation with the two earlier texts and through redirection and resignification opens the previous texts to different meanings that decenter colonial claims on history.

See Now Then centers on the lives of the Sweet Family and their breaking apart. Narrated from the perspective of the protagonist Mrs. Sweet, the novel examines the dismantling of her marriage over the course of several years. The text reflects on the interracial relations of Mrs. Sweet, who immigrated to the United States from the Caribbean as a young woman, and her Euro-American husband Mr. Sweet. This constellation mirrors that of the Creole wife and her English husband in *Wide Sargasso Sea* and *Jane Eyre. See Now Then* takes up these texts' discussions of interracial marriage, transporting them into the present by framing Mr. and Mrs. Sweet in the same structure of marital problems, and ultimately provides aspects of a female non-white perspective that remained absent in both *Jane Eyre* and (parts of) *Wide Sargasso Sea.*

The first of *See Now Then*'s eight chapters sets out to establish the relationship of its two protagonists Mr. and Mrs. Sweet. The focalization here alternates between the two characters in their thoughts about and descriptions of each other. However, ultimately the text conveys only Mrs. Sweet's perspective, as she is the narrator and fictional author of the book, so that her husband's thoughts about her would have been assumed by her. One of her descriptions of herself through her husband's eyes is particularly noticable for its repeated mentioning (eight times throughout the book, six of which occur on its first eighteen pages): she is not American by birth, but came to the United States as an immigrant, possibly having traveled by boat. This is a grievance of Mr. Sweet's. He perceives his wife to be very different from himself in all regards, which the novel presents as leading to the eventual end of Mr. and Mrs. Sweet's marriage.

Mr. Sweet regards his wife as of lower education and intellect, as is apparent in Mr. Sweet's dislike for a fine coat that Mrs. Sweet had bought for him from

85 See Sandra Gilbert and Susan Gubar, *The Madwoman in the Attic.*

86 Rhys claimed that she would attempt to write "[t]he real story – as it might have been" (qtd. in Pietrzak-Franger 11).

> a fine haberdasher in the city where Mr. Sweet was born and he hated the coat because his benighted wife had given it to him and how could she know what a fine garment it was, she who had just not long ago gotten off the banana boat, or some other benighted form of transport, everything about her being so benighted, even the vessel on which she arrived. (*SNT* 9)

Although she gifted the garment, he denies her understanding of its finery based on her heritage. The tripled use of the adjective "benighted" in this passage emphasizes Mr. Sweet's dislike for his wife. Beyond that, it points to his low opinion of her mental and educational development, as the term "benighted" entails converging meanings of backwardness, primitivity, darkness, and the absence of education in a Western sense.[87] The connection of darkness with these attributes might be read as pointing to underlying racist prejudices and misconceptions of people of color, which wrongly attest a perceived lack of learning to people based on skin color. The repeated use of the term "benighted" thus suggests Mr. Sweet's harsh dismissal of his wife's capabilities as racially charged. Mrs. Sweet is presented as thoroughly alien to her husband, further constituting Mr. Sweet as her opposite. He is aligned with the fineness of a coat from a fine haberdasher from the city he was born in, which establishes his heritage is as one of refinement.[88] Keeping in mind that Mr. Sweet's thoughts and judgments of his wife are conveyed by *her* in *See Now Then*, what the text ultimately presents are Mrs. Sweet's perceptions or possibly even imaginations of her husband's opinion of herself. She finds his reasons for disliking her and being unable to understand her, and hence for their incompatibility as life partners, in their racial and cultural differences based on her origins outside of the United States. Mrs. Sweet thus locates their marital difficulties within racial discourses of otherness and immigration as a threat to whiteness.

The historical dimension of these discourses is evoked by references to *Wide Sargasso Sea* and *Jane Eyre*, both of which also present marital relationships of white European husbands with immigrant wives from the Caribbean in which the unknowability of the perceived foreigner is at the center of the couples' inabilities to communicate. In *See Now Then*, Mr. Sweet reads one indicator for his wife's otherness in her eyes:

87 Synonyms for "benighted" include "analphabetic, ignorant, dark, nonliterate, rude, simple, uneducated, uninstructed, unlearned, unlettered, unread, unschooled, untaught, untutored" (*Merriam-Webster's Dictionary*).

88 Moreover, the phrase "banana boat" does not actually refer to a specific mode of transportation, but rather Mrs. Sweet's non-Americanness, as Mr. Sweet immediately concedes that she may have travelled by another means than a steam ship transporting both fruit and passengers. The phrase "banana boat" thus much more works as a discriminatory term to frame Mrs. Sweet as an immigrant from a country less wealthy – and in Mr. Sweet's opinion less enlightened – than the United States.

> Her eyes, dark, impenetrable Mr. Sweet would say, as he looked into them, at first he said the word impenetrable with delight, for he thought of discovering something not yet known to him, something that lay in Mrs. Sweet's eyes and that would make him free, free, free from all that bound him, and then he cursed her dark eyes, for they offered him nothing; in any case his own eyes were blue and Mrs. Sweet was indifferent to that particular feature of his. (*SNT* 19)

This passage echoes one from *Wide Sargasso Sea:* the novel is set in the first half of the nineteenth century and is concerned with the relationship of the Antoinette Cosway, the Jamaican-born daughter of a plantation owner, and an unnamed Englishman, who had traveled to the colonial island in search of wealth. He married Antoinette in order to secure her monetary assets for himself and thus allow himself to keep performing his social role of a genteel Englishman, which the laws of primogeniture jeopardized, as he was born as a second son. The marriage was meant to free him from his precarious condition of having to secure a substantial amount of capital not to embarrass himself and shame his family in not being able to maintain his social rank. However, the Englishman in *Wide Sargasso Sea* is disappointed when he ascertains, after pressuring Antoinette into the union, that she was not the right choice for him to achieve his goals of social acceptability. As a Creole whose family had been established in the Caribbean for generations, Antoinette already has a lower social standing in the Jamaican colonial society than more recently immigrated whites, as "the whiteness of white Creoles has become somehow culturally invisible when white Creoles are seen as others within whiteness" (Eeva 249). Observing his own wife more closely, Antoinette's husband at once sees indicators of racial otherness in and on her:

> I watched her critically. She wore a tricorne hat which became her. At least it shadowed her eyes which are too large and can be disconcerting. She never blinks at all it seems to me. Long, sad, dark alien eyes. Creole of pure English descent she may be, but they are not English or European either. (Rhys 37)

As in *See Now Then,* the wife's dark eyes are perceived as a marker of non-whiteness, which unnerves the English husband. It leads him to manipulate Antoinette in such a fashion that she becomes pliable to his wishes, for instance by calling her "Marionette" and eventually renaming her "Bertha Mason." Ultimately, Antoinette ends up in the attic of his English mansion, just as her literary ancestor Bertha Mason in *Jane Eyre.* At no point in *Wide Sargasso Sea* does the text present its female protagonist definitely as a person of color, or even as racially ambiguous. Rather, as different parts of the novel are conveyed from different perspectives by autodiegetic narrators, the text produces the Englishman's perception of his wife, which exposes the motif of the eyes as an external description that allows him to read race on her body.

Werner Sollors determines the description of eyes as dark and unreadable as a popular marker to identify literary characters of African descent, as for instance in Lydia Maria Child's "The Quadroons" (1842/46), William Wells Brown's *Clotel* (1853), and Harriet Beecher Stowe's *Uncle Tom's Cabin* (1852). Sollors demonstrates how the dark eye became a "familiar descriptive convention of the eye" in nineteenth-century literature (213).[89] Drawing on the "convention of the eye," *See Now Then* not only refers to *Wide Sargasso Sea*, but to a popular literary motif of racial otherness, which places the contemporary novel in historical contexts of literary interracial relationships.

However, *See Now Then* does not end with reproducing a technique of othering, but the contemporary text moves on to deconstruct it. In *Wide Sargasso Sea*, Antoinette Cosway's otherness is produced through her husband's perception and narration. While the text in general conveys the perspective of the marginalized Creole character Bertha Mason from *Jane Eyre*, here it remains mute, performing a silencing of the Creole wife. *See Now Then* responds to the motif of the dark, impenetrable eyes, when the contemporary text clarifies:

> But Mrs. Sweet's eyes were not impenetrable at all to anyone else and everyone she met wished that they were so; for behind her eyes lay scenes of turbulence, upheavals, murders, betrayals, on foot, on land, and on the seas where horde upon horde of people were transported to places on the earth's surface that they had never heard of or even imagined, and murderer and murdered, betrayer and betrayed, the source of the turbulence, the instigator of upheavals, were all mixed up. (*SNT* 19)

See Now Then here affords a change in focalization from the Euro-American husband Mr. Sweet and his racially prejudiced literary forebears to the experience of those who have been marginalized. In making the eyes readable, *See Now Then* presents them as windows to a history of slavery and colonization from the perspective of the abducted and abused. Redirecting the significations of the motif of the dark eyes from otherness to indicating the crimes of the colonizers, *See Now Then* reveals the turbulent history of oppression as equally inscribed in the motif of the dark as the other side of the same coin. It thereby highlights the plurality of histories as opposed to one History. The motif's previous meaning of unknowability and alienness is negated when the eyes are decoded, which not only

89 The Bertha Mason in Brontë's novel is also characterized by a dark appearance, with "a discoloured face," "black eyebrows widely raised over the bloodshot eyes" and a "blackened inflation of the lineaments" (372).

suspends the authority of this colonial motif, but also questions othering techniques based on physical markers in general. In this way, *See Now Then* takes up an intertextual "memory" of an earlier text (in Lachmann's sense) and revises it.

Although in *See Now Then* the marriage of Mr. and Mrs. Sweet ends in divorce, this certainly is a happier ending than that of Mrs. Sweet's two literary ancestors Antoinette Cosway and Bertha Mason in *Wide Sargasso Sea* and *Jane Eyre*. Famously, Bertha Mason dies in a fire of the English mansion, which in Brontë's novel sets her husband free of the burden of a mad woman for a wife. In Rhys's reworking of the material, Antoinette also suffers incarceration at the hands of her husband, who hid his Creole wife from English society. The novel ends with Antoinette dreaming of setting her husband's house on fire, which the text suggests she then does after waking (Rhys 123). In this case, it is Antoinette who is set free by the fire. This ending of the tragic characters of Antoinette Cosway and Bertha Mason is evoked in *See Now Then*, when Mr. Sweet muses that his wife "is strange and should live in the attic of a house that burns down" (159). Yet in this novel the fire remains a thought in the husband's mind. In repeating the character constellation and racial structures of the two earlier texts, *See Now Then* shows how Mr. and Mrs. Sweet's marital problems are far from unique,[90] and how nineteenth-century notions of interracial marriage affect contemporary imaginations of such relationships.

By framing its own characters in well-known narrative structures and evoking an established literary motif of othering, *See Now Then* visits the sites of colonial literary history, but in contrast to *Mr. Potter*, which highlights the effects of this legacy on contemporary textual production, the 2013 novel offers a counterperspective in resignifying the motif and thus decentering an objectifying white male gaze – both when it comes to concrete female subjects of color, and in terms of this gaze's overall interpretive authority. By taking up and continuing the motif of the "dark eyes" with a difference, the book challenges the totality of this motif through added meanings. *See Now Then* presents another take on history not as new, but as repressed but present – and needing only to be read. *See Now Then*'s intertextual engagement with *Wide Sargasso Sea* and *Jane Eyre* directs attention from the concrete events of individual narrations to their underlying structures of repetition, employing renarration to reevaluation and decenter colonial discourse's claims to interpretation.

90 Giovanna Covi similarly reads the main events of *See Now Then* as oozing ordinariness: "Falling in and out of love is ordinary. What has happened to Mr. and Mrs. Sweet is conventional. Kincaid implies that it does not require a particular explanation. Their story is a common story, not even worth the telling, and the narrator Mrs. Sweet concentrates instead on the changed perception of reality through time" ("Creolizing Cultures" 146).

Heracles and Persephone: Mythologies of the Ordinary

Jamaica Kincaid's latest novel *See Now Then* offers a wealth of intertextual relations from contemporary American literature to children's books to popular culture.[91] The novel's allusions to Greek mythology nevertheless stand out for their ubiquity. For instance, the Sweets' son and daughter are named Heracles and Persephone. Both characters also display features of their mythological counterparts, which makes for extraordinariness in the rather ordinary world of the Sweet family. The exceptionalism thus displayed in *See Now Then* starkly contrasts with the pervasive ordinariness in *Mr. Potter*, Kincaid's latest novel before this one. To *See Now Then*'s homodiegetic narrator, Mrs. Sweet, much of the world and especially her children are a marvel. This, I argue, is expressed through intertextual engagements with Greek mythological material. Within the renditions of Heracles and Persephone's lives, the extraordinary is moreover produced in the text's play with literal and figurative language, which diffuses meanings and continually lets the text spin off into excess and exaggerations that create both proximities and distances with its contents. In a motion of approximation and distancing, the text positions the banal as grand and the extraordinary as trivial. Ultimately, this back-and-forth motion between the exceptional and the conventional, created through intertexts and excess of detail, inscribes a nuclear American family into contexts of Greek mythology, which presents their daily lives as equally fabled as those of Greek heroes and heroines. At the same time, the text maintains the ordinariness of even the most exceptional circumstances and events, which in turn imbues the famous characters of Greek myth with human banality.

See Now Then spans the married life of the Sweets and the maturing of their two children from birth to their teenage years. The narrative is set around or just before the turn of the twenty-first century in the family's home in the small town of Bennington, Vermont. The description of the Sweets' house already contains references to other literary texts, as throughout the novel it is called the "Shirley Jackson house" (see for instance *SNT* 3, 21, 49, 72, 109, 135, 162, 182), named so for a previous resident of the home. Following up on the invocation of Shirley Jackson, a wealth of intertextual links between *See Now Then* and Jackson's own work reveals itself. The reading motion incited by the cue "Shirley Jackson house" at the very outset of the novel is exemplary for the play with intertextual cues. In many cases, the book incites paper chases through text after text that eventually lead

91 I am here referring, for instance, to Shirley Jackson (whose mention in *See Now Then* I will also discuss on the following pages), the mentions of Beatrix Potter's *Peter Rabbit* (*SNT* 124), Margaret Wise Brown's *Goodnight Moon* and *Runaway Bunny* (*SNT* 113), and the hip-hop duo Outkast (*SNT* 158).

to the simple realization that text generates further text. In this, *See Now Then* perfectly exemplifies the productive aspect of intertextuality that Renate Lachmann emphasizes: the intertextually organized text "is produced by process[es] of reference to other texts" (30). *See Now Then*'s engagement with Shirley Jackson demonstrates the text's general mechanisms of its intertextual work. Hence, and because the Shirley Jackson house is itself entangled with the Greek intertexts based on its architecture, I will briefly focus on Jackson, before I attend to the Greek mythological intertexts and their links with a conception of history in Kincaid's latest novel.

American author Shirley Jackson lived and died in the small village of North Bennington, Vermont. She and her family resided in several houses here, all rented except for the last one. The first of these is located on 12 Prospect Street, "an 1850 Greek Revival mansion complete with two-story columns" (Petrie). Jackson's last North Bennington home, where she settled permanently in 1953, is located on 66 Main Street, which according to her biographer Ruth Franklin is still known in the Vermont town as "the Shirley Jackson house" (326). Kincaid in *See Now Then*, however, chose to give the moniker to Jackson's first home, which is the one that also found entry to Jackson's own literary works, as the covers of her nonfictional work *Life Among the Savages* suggest. For instance, the first edition (Farrar, Straus and Young, 1953) or the currently easily available edition by Penguin Books (1997) feature the white columns of the house's front porch in their front cover art. The Prospect Street house arguably is the setting of Jackson's autobiography *Life Among the Savages* in which she describes the family home as "modeled [...] after, presumably, a minor Greek temple; [...] [with] four massive white pillars across the front" (11). Jackson assumes the reason for the architectural choices in the "classical revival [...] upon the country" when the house was built in roughly the 1820s (11). Enthusiast for writer's homes Kathye Fetsko Petrie regards the Greek revival mansion on 12 Prospect Street as the setting not only of *Life Among the* Savages, but also of the later *See Now Then.* This is easily supported by *See Now Then,* which depicts its own setting as a "house with great big Doric columns, Victorian and Greek revival architecture" (*SNT* 135–136). The location of the house is moreover mentioned so specifically that it is easy to follow Mrs. Sweet on a map of Bennington, when she drives to pick up her children from a school bus stop. On their way back, they start out from Bennington Monument (*SNT* 127), driving along Silk Road, crossing the Walloomsac River, along Matteson Road and onto Harlan Road (*SNT* 131), which merges into Prospect Street.

Yet all this detective work of following the clues in *See Now Then,* which leads to the biography of another American author and a map of her home town, hardly provides deeper insights into *See Now Then* itself, except the real-life location of a house that serves as a setting for a fictional text. Further reading up on Shirley Jackson's life reveals remarkable parallels to Mrs. Sweet's own biography: both their husbands – Stanley Hyman and Mr. Sweet – taught at Bennington college, both are of Jewish heritage, both their families were against the unions with non-Jewish wives, and both men left their wives for younger women, specifically their college students. This repetition of personal living circumstances of two women writers in a singular space but at a distance of roughly fifty years is expressed in the very similar narratives set in the space of the Prospect Street house: the peculiarities and banalities of domestic life with children.[92] Moreover, it points to an absence of originality in human biographies, which is one of the main uses of intertextuality in *See Now Then.* Beyond that, such paper chases, which can be vastly entertaining to the willing reader, quickly exhaust themselves in laying bare only trivial connections, but they do not elucidate the intertextual relations. *See Now Then* continually makes such offers in suggesting external texts, which ultimately only lead back to the everyday lives depicted in *See Now Then.* For instance, the family car is a "Kuniklos" (*SNT* 103 and 134). The Greek name of the car might appear significant in the context of other Greek references in *See Now Then.* A hunt for meanings will eventually reveal that the German car manufacturer Volkswagen produced a model called Rabbit between 1974 and 1983. Outside the United States, this model would be known as a first-generation VW Golf. "Kuniklos," of course, is Greek for "rabbit." Coming back to *See Now Then* with this added knowledge does not open Kincaid's text to further meaningful significations, but this is a return to an ordinary family driving an old car.

Just like the Jackson-Hyman family, the lives of the Sweets are ordinary ones. *See Now Then* illustrates this with a myriad of ostensibly unimportant details, such as how Mr. Sweet likes his coffee (Maxwell House instant coffee with creamless milk, *SNT* 72), the son's favorite toys (McDonald's Happy Meal gifts, *SNT* 24 or 49), or Mrs. Sweet's knack for home-cooked French food (*SNT* 7). Descriptions of the family life center on domestic worries and duties, such as the disturbing noises of the washer and dryer (*SNT* 22 or 164) or due payments of car loans (*SNT* 61). The banality of the Sweets' life is set within the ordinariness of small-town life. Here, neighbors help each other out, such as Homer cleaning out the gutters of the Sweets' roof (*SNT* 4).[93] The Sweets' neighbors, too, are generally characterized by their ordinariness. They wear the same pairs of socks for several days and worry about their boilers and pipes freezing during the winter (*SNT* 7–8). The so-

92 Apart from this overlap, Shirley Jackson's and Mrs. Sweet's literary works differ considerably, as Jackson is best known for her gothic novels and short stories, whereas Mrs. Sweet (like Jamaica Kincaid) is most interested in personal and colonial history (*SNT* 145).

93 On one of the novel's first pages, the Greek writer Homer is evoked, if only by association when the text mentions the neighbor's first name. Nonetheless, this sets the mood for the Greek intertexts throughout the book.

phisticated Mr. Sweet is angered by their falling asleep during his piano performance of Shostakovich (*SNT* 6–7), which leads him to despise his audience wearing "duck-feather-filled coats, which had trapped the smell of wood smoke" (*SNT* 8), with "their Subarus and second-hand Saabs" (*SNT* 9).[94] The descriptions of inconsequentialities, such as the socks worn more than once, or Mr. Sweet's preference for burnt toast (*SNT* 72) thoroughly cast the Sweets and their neighbors in the ordinariness of rural small-town life.

This very profane world becomes a marvel through its narration, when Mrs. Sweet, "would find the simplest thing extraordinary" (*SNT* 87). Her references to Greek mythology seemingly transport extraordinariness into the ordinary. This is already evident in the children's names Heracles and Persephone, contrasted by their playmates' nondescript names Tim or Tad (*SNT* 96 or 149). The majority of *See Now Then* centers on the second-born child of the family, Heracles,[95] and his growing up from a nursing baby to a toddler to a boy in his preteen years. Heracles is characterized by his youthfulness (*SNT* 36) and physical prowess (*SNT* 107). He is tasked with different duties and has to weather perils that insert themselves into the life of the toddler. For instance, he is woken up by his sibling's yells while asleep in his crib as an infant:

> Wake up, his sister shouted at him, a snake with nine heads is lying next to you in your crib. And the very young Heracles then turned over into somersault, and facing the nine-headed snake directly stuck out his tongue at all those heads; without making too much of an effort, he tore off their heads and threw them over his shoulder, all nine of them. (*SNT* 45)

The danger of the snakes and the victory over them reminisces the fabled Heracles's defeat of the serpents sent by the jealous Hera to kill the extramarital son her husband Zeus begat with the human woman Alcmene,[96] which Pindar tells as follows:

> but the queen of gods with anger in her heart immediately sent snakes. When the doors had been opened they went into the deep recess of the bedroom, eager to wrap their darting jaws around the babies [Heracles and his half-brother Iphicles]. But the boy [Heracles] lifted his head straight up and engaged in his first battle, grasping the two snakes by their necks in

94 Were Mr. Sweet's biography written, it might well use the same title as Shirley Jackson's, *Life Among the Savages*, for his dislike of Bennington's inhabitants.

95 Mrs. Sweet, the narrator, rather focuses on Heracles than on her first-born daughter Persephone. Persephone is portrayed to be her father's daughter, Heracles as his mother's son.

96 Son of a human and a god, the Greek Heracles is born a demigod. As the realms of ordinary contemporary small-town life and myth are blurred by intertextual references to extraordinary feats of a small boy, so are the realms of human and divinity in the body of Greek Heracles.

> his two inescapable hands, and as they were being strangled, the passage of time exhaled the life from their monstrous bodies. (Nemean Ode I, 9–11)[97]

However, while in Pindar's rendition of the myth, Heracles's slaying of the snakes establishes his "extraordinary determination and power" to the fellow human beings of his world (to his adoptive father Amphitryon; Pindar 11), in *See Now Then* it does not. Here, the snake heads end up in an ordinary world which takes little notice of the infant's feat and which at most is inconvenienced by the outcome of this event: the organic waste produced in Heracles's fight very ordinarily lands "on the floor of Mrs. Sweet's newly cleaned kitchen" (*SNT* 45). Unimpressed by her son's victory and his averting of a deadly threat, Mrs. Sweet is more concerned with mundane household duties: "Oh god, she said to herself, that kid is always up to something, what a mess he has made now. And she picked up the nine snake heads and put them in a bag, wiped the floor clean, and asked Mr. Sweet to please come and put out the garbage" (*SNT* 45). The narrative immediately returns to normality after the serpents' deaths by removing the disruption and thus reverting the domestic space of Mrs. Sweet's kitchen to its previous state sans snake heads. While Heracles's achievement marks his exceptionality in the Greek myth, the contemporary text refuses to recognize this. The intertextual reference to Heracles's deed might fuel expectations of proclamations of heroism, but the return to Mrs. Sweet's kitchen disregards such implications. Marveling at Heracles's abilities is left to the reader.

See Now Then operates similarly when it comes to mythical Heracles's labors, which are intertwined with tasks of the young contemporary Heracles's (everyday) life:

> wash the dishes, put them away, clean the stables, walk the horses, fix the roof, milk the cows, emerge from his mother's womb in the usual way, slay the monster, […] lay waste to whole villages to the surprise of the villagers, trap and skin the she-fox, eat his green vegetables and his meat too, […] cross the road by himself, tie his shoelaces, kiss a girl, sleep in his own bed. (*SNT* 36–37)

This list may be read as the labors of growing up, interspersed with metaphorical renderings of child's play, such as the destruction of toy figurines lined up in toy villages. When Heracles grows older, the mythical references become more pro-

---

97 See also Apollodorus's *Library:* "When the child was eight months old, Hera desired the destruction of the babe and sent two huge serpents to the bed. Alcmena called Amphitryon to her help, but Hercules arose and killed the serpents by strangling them with both his hands" (Book II, Ch. ix, 175). Moreover, *See Now Then*'s Heracles kills a nine-headed snake, which additionally references the mythical Heracles's defeat of the Lernean Hydra.

nounced – as do the real-life dangers he encounters. The adventures his mother foresees in his life

> might be cleaning the fabled Augean Stables, slaying the Nemean Lion and wearing his skin as a cloak, an encounter with the Erymanthian Boar, though not yet and perhaps never the policeman in the city of Boston, who traces himself through some long-dead people from Ireland, imagines the young Heracles has run through a red light and by that time, then and now, the young Heracles had become a young black man, whatever that might be, and even now, whatever that might be is not certain. (*SNT* 104)

The link of mythical Heracles's feats with the possible encounter with a policeman in contemporary Heracles's life suggests that being stopped by a white police officer is as perilous in the life of a young Black man in the United States (here Boston) as mythical Heracles's labors.[98] In turn, this implies that the survival of such an encounter would be as marvelous as the supernatural Heracles's victories over powerful beasts,[99] which exposes everyday tasks of African American men to be as perilous as the ordeals of a demigod.

The portrayal of Heracles's older sister Persephone similarly takes up the narrative of her mythical prefiguration: most noticeable about Persephone in *See Now Then* is her absence. Persephone hardly emerges as a defining presence in her mother's life or in her narrative. Mrs. Sweet attributes this to the efforts of the girl's father, who "[i]mmediately after Persephone's birth, [...] began to secrete her" (*SNT* 123). At times, Persephone is kept in Mr. Sweet's study (*SNT* 59), which is associated with death. To Mrs. Sweet her husband's private space "looked like a replica of a welcome area of a funeral home" (*SNT* 14). At other times Persephone is "carefully hidden in his pocket, out of her mother's sight" (*SNT* 79) from which she may emerge in spring (*SNT* 128–129). Persephone's absence, her associ-

---

98 Apollodorus's *Library* counts slaying the Nemean Lion, catching the Erymanthian Boar, and cleaning the Augean Stables among the fabled Heracles's labors, which he has to complete in order to earn immortality (see Apollodorus 185–237).

99 Kincaid and Mrs. Sweet point to the general perils of living as a young Black man in the United States, where traffic offenses might be imagined by police officers on the basis of skin color. This is as relevant as ever in the 2010s, which recorded the deaths of men including Eric Garner, Michael Brown, and Tamir Rice that have become synonymous with excessive and unwarranted police force against Black men and boys in the United States. The proximity of the Greek Heracles's fabled labors and the young Black Heracles "adventures" thus also conveys a critique of contemporary social and political circumstances – in particular the contemporary life-threatening racism and brutality of law enforcement officers towards young Black men in the United States. Through its references to Greek mythology, *See Now Then* suggests that in order to survive day-to-day in the United States, Black men need to be superhuman.

ation with spring,[100] and her dwelling in a space associated with death recall the mythical Persephone's residence in Hades's underworld, from which she may travel during spring time.[101] The mythological intertext motivates Persephone's absence and the unknowability of her character.

Tracey Walters maintains that "the Persephone and Demeter myth is by far one of the most popular myths adopted by Black women writers" (19). Refigurations of the myth, according to Walters, often serve to highlight Black women's precarious situations by focusing on the rape and abduction of the mythical Persephone (27). Moreover, the myth is often revised in such a way that Persephone's perspective is represented.[102] Such a recovery project of a hitherto muted female voice is decidedly not at stake in *See Now Then*. Here, Persephone remains quiet and unknowable. In this way, the character remains a type,[103] mainly defined by character constellation, i. e., by her absence from her mother.

Since the diegetic world of North Bennington is a very profane one, it might be clear to the reader that Persephone does not fit into Mr. Sweet's jacket pocket at all, for she is a grown human girl, but the text here does not recognizably mark Persephone's living in a jacket pocket as metaphorical. Supernatural circumstances, such as conflicting sizes, are accepted as facts by the text. It pays no further attention to them. Here, the impossible is possible,[104] and the extraordinary mundane.

The mention of Persephone's living inside her father's jacket pocket is moreover embedded in a wealth of detail: the scene begins with Persephone and Heracles waiting for their mother to be picked up at a bus stop. Persephone's presence in this scene is specifically marked, which yet again emphasizes her general absence: "it was spring and she was released from living in the depth of the jacket of Mr. Sweet's old Brooks Brothers tweed jacket (and the lining of that pocket was made of silk purchased in Hong Kong)" (*SNT* 128). The second part of this quote doubly specifies the materiality of the pocket. Firstly, the jacket itself is de-

---

100 Persephone not only returns in spring, the text even equates her with spring itself: "the young Persephone was like a season, spring – just for instance – and spring especially!" (*SNT* 147).

101 See "Hymn to Demeter," lines 441–459, and Apollodorus, Book I, Ch. V, 41.

102 Walters here refers to works by Gwendolyn Brooks, Toni Morrison, and Rita Dove.

103 Reading Heracles and Persephone as character types is moreover supported by their names almost exclusively occurring in compounds: "the young Heracles" and "the beautiful Persephone" (see *SNT* 3, 24, 43, 66, 90, 133, 173, 180, etc.). The narrator even bespeaks a permanence of the attributes of youth and beauty, which in the text at hand is manifested in the adjectives used as parts of the names: "the young Heracles who would always be so, not matter what befell him, and the beautiful Persephone, who would always be so, beautiful and perfect and just" (*SNT* 13).

104 The same literary technique is at play in Mrs. Sweet's knitting a pair of testicles from the heels of two woolen socks when her husband loses his original organs in an altercation with their adolescent son Heracles (*SNT* 46–47).

fined as being made of tweed and as having been manufactured by an established American gentlemen's outfitter. Secondly, the parentheses define the consistence of the pocket lining, which again emphasizes the material's exclusivity. While these details (again) establish Mr. Sweet as a man of particularly American sophistication, they do not speak to the topic at hand – Persephone's superhuman (changes in) size. Rather, these details disregard the exceptional aspects of Persephone and focus on very mundane aspects of the jacket.

A few lines further, Persephone seeks to comfort her brother who is distraught at their mother's absence and "[n]ot knowing what else to do, she lifted him up with much ease, [...] and she placed him in the right-hand pocket of her own jacket which was made from polyethylene terephthalate and the pocket itself was lined with rayon" (*SNT* 128). Persephone here repeats her father's concealing of the familial body. This passage also repeats the previous one in that it conveys details about Persephone's jacket pocket. The material polyethylene terephthalate is a specific type of polyester, better known by the abbreviation PET. As with the silk in Mr. Sweet's pocket, the detail of the exact material of Persephone's jacket does not add to the narrative except in disrupting the sentence with its two multisyllabic words from chemical terminology. Again, detail here obscures the extraordinariness of the events. Either Persephone must have instantaneously grown to enormous heights – and her apparel with her – or her brother just shrank to the size of a hamster in order to fit into Persephone's pocket. The text itself does not call attention to this unlikeliness, which normalizes the extraordinary, or – the other way around – turns the normal into the exceptional.

Ultimately, although mythical intertexts at first glance appear to intersperse *See Now Then* with exceptional, superhuman, and awe-inspiring moments, they do not. The supposed extraordinariness is inserted so seamlessly into the text of *See Now Then* that it loses its specialness. Generally, utter ordinariness prevails in the Sweets' home life and the threat of a nine-headed snake is of less interest to the characters than the inconvenient mess its death produces. In this, when *See Now Then* takes up names and characteristics of Greek myth, the contemporary text disregards their perhaps most important features – their exceptionality. Heracles and Persephone remain types, defined in part by their fabled actions, such as the labors in Heracles's case, or their character constellations, as with Persephone. Both characters do not develop psychological depth or individual character traits. Such "types" in conjunction with pervasive ordinariness uncomfortably recalls the general condition of Mr. Potter in Kincaid's previous novel. Here, as I am arguing, the ordinariness finds its cause in the hegemony of Western colonial history that does not allow for imaginations of a descendant of enslaved Africans as an evolved and well-rounded human being. *See Now Then* presents the same constellation of ordinariness, albeit in a different situation: a nuclear American family is depicted as ordinary not through colonial intertexts, but by reference to Greek myth.

I contend that by taking up Greek mythology, *See Now Then* engages with the roots of Western history. Édouard Glissant locates the shared origin of literature and history in Greek myth. "Myth is the first state of a still-naïve historical consciousness, and the raw material of the project of a literature," Glissant writes (*Caribbean Discourse* 71). *See Not Then*'s engagement with myth may thus be read as a turn to European ur-history in literature, into which the Sweet family now is integrated. Their different historical status is especially evident when considering Mr. Potter, who was of no account and thus excluded from history. The Sweets, however, are presented as embedded in much larger contexts that are the antecedents of colonial history.[105] The association of the contemporary characters with mythical figures presents them as universal types that are ordinary because of their humanity. The repetition of characters constellations and structures presents ordinariness as a ubiquitous human condition. The totality of colonial history cannot serve Kincaid's characters, but stepping beyond that into literary history, the obscurity of myth can. As *See Now Then* renders the heroes of Western myth as ordinary as an American family of four, the text strips them of their extraordinariness.[106] Ultimately, in this, *See Now Then* decenters the totality of Western exceptionalism both in literature and in history.[107]

---

105 Barbara Goff makes the connection between the classics and colonialism clear: "Latin and Greek language and culture were so inseparable from the elite's vision of itself that they become inseparable from the vision of the imperial role [...]. The role of classics in elite culture also predisposes it [...] to be closely connected with imperial projects: because classics bears with it the weight of tradition and authority, it can easily be pressed into service as a *sign* of tradition and authority in general" (11, emphasis in original). Hence in picking up Greek myth, *See Now Then* negotiates the prototypes of imperial writing.

106 In this, *See Now Then* is reminiscent of Derek Walcott famously transferring Greek mythology and history to the Caribbean in his epic poem *Omeros* (1990), which renarrates ancient text through more recent and still ongoing West Indian history with a focus on contemporary and especially ordinary characters: "Forget the gods, [...] and read the rest," Walcott's poet Omeros proclaims (283).

107 Lorna Hardwick also points out that "classical texts have been displaced from their perceived centrality in Western culture, [because] they have become part of a diaspora – migratory, open for interaction, yet enabling an awareness of their own tradition in the ancient world, which is increasingly understood as involving a multi-valent political discourse" (116).

2.3 Writing between and across Genres: Negotiations of Subject and Object, Self and Other

Jamaica Kincaid's work ranges from short texts, published in magazines (foremost *The New Yorker*),[108] to 200-page novels, and from descriptions of family life to travel writing. Kincaid collaborated with visual artists Eric Fischl (lithographs, 1986), Lynn Geesaman (photography, 1999), Ricardo Cortés (illustrations, children's book, 2019), and she worked as an editor for three anthologies (1995, 1998, 2005). Most of Kincaid's works are commonly regarded as being situated between autobiography and fiction. Likely for their recurring autodiegetic narrators, Kincaid's texts are often understood as fictionalizations of the author's own life. Kincaid explicitly rejects the concept of generic categories: introducing an edited volume of essays, Kincaid's first sentences reproach the genre that is at the center of the collection:[109]

> An Essay! The fixed form or fixed category of any kind, any definition at all, fills me with such despair that I feel compelled to do or be its opposite. And if I cannot do its opposite, if I can in fact complete the task that is the fixed form, or fill the fixed category, I then deny it, I then decline to participate at all. Is this a complex view? But I believe I have stated it simply: anything that I might do, anything that I might be, I cannot bear to have its meaning applied to me. (xii)

Kincaid then goes on – in this introduction to a collection of essays – to criticize the genre "essay" as a form that commonly expresses opinions and observations some people have on others, which "can only, ultimately, fix you, categorize you" (xii). Categorization and fixation as aspects of colonial world ordering thus align the essay in particular, and genre as textual categorization in general, with colonial thought. Nonetheless, Kincaid states that she finds pleasure in some such texts and posits her own idiosyncratic pleasure principle as the only criterion in the assembly of the following collection (xiii).

I propose that rather than musing about a "true-to-life-veracity" of certain auto/biographical aspects of her texts,[110] or about her simultaneous dismissal and enjoyment of the essay form, we could understand Kincaid's back-and-forth between generic expectations as exemplifying what her texts perform – a questioning of *genre* itself. In "doing" genre differently, Kincaid's texts both investigate the histories of generic categories and present the inability of generic fixity to rep-

---

108 See many of them collected in *Talk Stories* (2001).

109 In 1995, Kincaid was the guest editor of Houghton Mifflin's The Best American Essays series.

110 See, for instance, Daryl Cumber Dance's meticulous research into the lives of the author and her family.

resent Caribbean colonial experience. The subversion of generic conventions thus opens them up to the voices of Kincaid's protagonists. Kincaid's poetics of impermanence dissolve generic fixity when the established form (that of the essay, the autobiographical, etc.) is both rejected and appropriated, when it is claimed for oneself and thus utilized to disintegrate the power of the colonizer/canonizer, and to claim one's own power to speak. Kincaid's works seem "to be concerned to critique and revise the modes of representation fundamental to the canonical texts" that make up traditions of writerly expression in the sense of Henry Louis Gates's Signifyin(g)[111] (*Signifying Monkey* 217). In Kincaid's case, these are conventions of the autobiographical, biographical, and the fictional that are signified on. I argue that in Kincaid's works structures of repetition enter these generic concepts into conversations with each other. Generic forms of writing are brought together and intermingle in a seemingly playful relationship. Ultimately, this refigures established forms, questions them, and utilizes them to comment on the colonial complicity of both autobiography and biography as forms of self-narration. By Signifyin(g) on them, and by thus laying bare their mechanisms of framing "the colonial subject" in a position from which it (or rather: he or she) cannot speak, Kincaid's works and especially *The Autobiography of My Mother* and the "Biography of a Dress," which will be at the center of the following analyses, interrogate genre conventions to negotiate the emergence of a speaking subject from an art form (writing) that has historically objectified the colonized. Through their engagements with genre conventions Kincaid's texts renegotiate the (post)colonial speaking subject.

In the following observations on Kincaid's texts' signifying on genre, I do not regard genres as fixed categories themselves. Rather, I grasp "genre" as a learned category that, as a reading strategy, imposes fixedness on text. Tzvetan Todorov writes on the origin of genres that "[i]t is because genres exist as an institution that they function as 'horizons of expectation' for readers, and as 'models of writing' for authors" (163). Just as expectations and institutionalized categories frame the reception of text, this reciprocates in texts in that they "participate" in genres, as Jacques Derrida puts it. Derrida describes the relationship between text and genre as a "participation without belonging – a taking part in without being part of, without having membership in a set" (59). As such, genre "takes place" both in the text itself as well as in the modes of its production and reception. Kin-

111 Gates phrases this with specific regard to Ishmael Reed's novel *Mumbo Jumbo* and its Signifyin(g) on the "tradition of the Afro-American novel," which comments "on the history of the black novel" (217). I take up Gates's wording to extend it beyond exclusively African American concerns to Kincaid's critique of "Western" colonial canonical texts in a Caribbean context.

caid's works, as I show, actively perform genre in order to highlight and subvert reader expectations.

Jacques Derrida notes that "the law of genre" is "a principle of contamination, a law of impurity, a parasitical economy" (59). He furthermore emphasizes the vital "'anomalies' [that] are engendered [...] by *repetition*" (57–58, emphasis in original). Tzvetan Todorov refers to the "laws" of genre to describe the individual genres' categorizing principles (160). A text's nonconforming contamination in this conception then does not make genre "nonexistent; it is tempting to say that quite the contrary is true. And for a twofold reason. First, because transgression, in order to exist as such, requires a law that will, of course, be transgressed. One could go further: the norm becomes visible – lives – only by its transgressions" (Todorov 160). I regard such transgressions in Kincaid's texts as ostensibly performed for the purpose of commenting on genre's forceful aspect of categorization, which imposes meanings on texts, just as the colonizer imposes meanings on the colonized. At the same time, Kincaid's texts also relish in the productive aspects of contamination when they rework aspects of the genres they invoke to create space for the voices of the colonized, which in these genres historically are fixed in the positions of the object or the voiceless.

In *The Autobiography of My Mother* and "Biography of a Dress," as the titles of the two works already suggest, the genres in question are autobiography and biography. Along with that, *The Autobiography of My Mother* is denoted as "a novel" on the book's cover, and "Biography of a Dress" suggests fictionality in the unlikely writing of the life of an inanimate object. Thus, both texts bring forms of life writing into conversation with fictionality. As I demonstrate, Kincaid's texts repeat conventions and thus transform them into something else. Yet I do not reconsider the genres of autobiography, biography, and novel themselves to finally devise a new term to describe Kincaid's alterations to existing categories and to grasp them as a new (postcolonial) subcategory to established forms, as for instance Jana Evans Braziel or Leigh Gilmore do.[112] Rather, I focus on the texts' mechanisms in engag-

112 Leigh Gilmore considers Kincaid's oeuvre as "serial autobiography," with each new book adding a volume to the series ("What Was I" 100). "Serial autobiography permits the writer to take multiple runs at self-representation, more as a way to explore the possibilities present within autobiography than to produce a single, definitive solution to the problem of presenting identity" (*Limits* 103). This practice of seriality, in Gilmore's view, questions the concept of autobiography itself. It "challenges the limits of the genre by raising the specter of endless autobiography. That there will always be (another) autobiography means that there will be no last words and autobiography is a genre of last words" (*Limits* 96). This multiplication of self-representation thus transforms a key aspect of the genre in that it fragments the unity of Western autobiography by defying its finality. Jana Evans Braziel develops the notion of "alterbiography" with reference to Kincaid's works. Braziel compounds her neologism from "alter-" as in alterity and "biography." The focus on

ing with genres and their conventions to show how textual strategies, such as signifying on genre conventions, purport to perform genre (especially autobiography) in order to subvert them. Heteroglossia, as theorized by Mikhail Bakhtin, creates a simultaneity of voices and versions of the self that question the objectification of the colonized and make space for speaking subject positions, or multiply the voices of those spoken for in a self-narration through an other.

The first section of this chapter, "Beyond Auto-Bio-Graphy," focuses on *The Autobiography of My Mother* in which a debate of the genres autobiography, biography, and the novel plays itself out in a tension between an "I" and a "you." The narrative of the singular self is opened up to the voices of a multitude, contained within the "I." *The Autobiography of My Mother* engages in discussions of genre to present the lasting effects of colonial rule on a collective and demonstrates how generic conventions themselves participate in colonial discourse. The second section of this chapter attends to Kincaid's shorter work "Biography of a Dress" and explores how the writing of the self here is entangled with the history of the eponymous dress, as well as a soap advertisement, which brings with it colonial discourses of beauty, purity, and girlhood. Historically once also a commodity, the body of the Black girl protagonist is spoken about by a narrating self, temporally removed from the experience of objectification. I argue that through the tensions between autobiography and biography, "Biography of a Dress" debates subject and object positions available to its girl protagonist and adult narrator.

Beyond Auto-Bio-Graphy: Writing the Self Through an Other

Kincaid's 1996 novel *The Autobiography of My Mother* uses genre conventions to explore who has the agency to speak of oneself by exploring who might speak for or with whom. This section interrogates the engagement with established generic forms in *The Autobiography of My Mother* in order to examine the involvement of genre in the writing of both personal and collective history. *The Autobiography of My Mother* appears to adhere to the form of classic autobiography. Yet it is marketed as a novel. I read it as a fictional text that virulently offers autobiographical markers to its readers only to withdraw itself from fully being read as such.

---

alterity as in otherness to the typical Western autobiographical subject as well as on an Other as part of the constitution of a self is consistent throughout Braziel's very useful analyses of Kincaid's works, as it points to the transcendence of genre boundaries, especially between autobiography and biography. Likewise, Alison Donnell, in her article on *The Autobiography of My Mother*, initially grasps Kincaid's work with Liz Stanley's notion of "auto/biography," which similar to Braziel's alterbiography insists on the confluence of the lives of self and other.

The text appears to feed the readers' willingness to assume the existence of the narrator and protagonist in their own real world, as well as a truthfulness to the report of her life. In simultaneous offering and withdrawing of assumptions of authenticity, *The Autobiography of My Mother* performs different aspects of genre conventions without indeed committing to them. This confusion comments on the nexus of subject and colonial history: the lives of the text's characters are all inextricably bound up with the colonial history of the Caribbean islands of Dominica and Antigua. Written in the first person, a seemingly daughterly "I" narrates the life of a mother figure, which creates a tension between the writing self and the life that is written, i. e., between the *autos* and the *bios* in the term "autobiography." Moreover, while an "I" is at the center of the text, and while the protagonist Xuela uncompromisingly positions herself as the narrator, this "I" splits into a multitude of voices. History and genre thus are negotiated between self and other, which this section investigates by asking the following questions: Who speaks? Who is spoken about? How do these acts of speaking engage with categories of genre? Finally, how do its generic interventions speak to the text's engagement with colonial history?

Several aspects indicate the autobiographical mode in this novel. *The Autobiography of My Mother* explicitly bears the genre designation in its title. An autodiegetic narrator relates the life of Xuela Claudette Richardson, who also is the text's protagonist, in retrospective from infancy to old age. In his seminal work *On Autobiography*, Philippe Lejeune posits that to qualify as autobiography, a work must have an author, narrator, and protagonist that can be identifiable as identical. In *The Autobiography of My Mother*, narrator and protagonist are explicitly presented as identical by the identity of their names. While the author name Jamaica Kincaid does not participate in Lejeune's assumed triad of identity, this lack is taken up by the title's suggestion of the autobiographical genre, which appears to offer the "autobiographical pact" in Lejeune's sense that advances the assumption of truthfulness in the reader's perception. The title of the book, however, also questions key suppositions of the genre. The pronoun "my" in the book's title points to a first-person speaking position. At the same time, it refers to "mother," who through the possessive preposition "of" is posited as the author – or at least owner – of the autobiography. This stands in opposition to the *autos* in "autobiography," which also requires an autodiegetic narrator. The title thus already contains two autodiegetic positions. Moreover, the book title, as printed on the cover of the first edition seemingly continues with "by Jamaica Kincaid." Both printed in handwriting, the title and author's name are not visually differentiated from each other. Rather, in its connection with the title through the preposition "by" it could be read as part of the book's title. This adds another entity to the list of possible speakers, namely that of a real-life author, which suggests that the speaking self might be that of Jamaica Kincaid and that the mother in question might be Annie Drew, the author's own mother. Such an assumption, however, is swiftly confounded by the addition "a novel" in the lower right corner of the cover. The claim to the fictional genre immediately discredits any inference about author and narrator. In all of this, the book's title already poses the questions "who is speaking about whom?" and "how do these speaking positions relate to each other?" These issues shall be discussed in detail later in this section, after attending to the general problem of speaking of the self in a (post)colonial context, which *The Autobiography of My Mother* showcases.

To begin with, the novel lays bare the lasting effects of colonialism and slavery on its (former) subjects by denying their ability to speak of and for themselves. This plays out in the negotiation of genre as a colonial category of classifying textual production. As a lasting effect of European colonialism, the social position of the novel's protagonist Xuela is precarious. She is an Afro-Caribbean woman who may visually be identified as the descendant of both African people and Dominica's native Caribs. Xuela understands herself as belonging to the "defeated" (*ABM* 215). With this term, she refers to the history of enslavement and colonialism in the Caribbean.[113] History thus determines Xuela's contemporary position, which through her mixed heritage is doubly precarious: she describes the Dominican people of African descent as "defeated," but worse than that she sees herself as ostracized for her Carib ancestry. It is looked down on by other Dominicans as "exterminated, thrown away like weeds in the garden," while people from Africa at least "survived" the destruction of the colonizers and slavers (*ABM* 15–16). This historically conditioned circumstance, Xuela argues, positions her within the realm of the "unreal." With this, she frames the lives, experiences, and thoughts of those who have been "defeated" from the viewpoint of the victor. Xuela's definition of her position is thus informed by the entangled discourses of history. The victors determine what is "real" and what is not. "Reality" in this sense is far from what people experience. This precondition of a separation of reality and experience already problematizes the speaking of oneself. If autobiography is supposed as the truthful account of one's own past life, it would be impossible for Xuela, as one who has been "defeated," to participate in this genre, since her experience is discredited as insubstantial and fictitious from the start.

In *The Autobiography of My Mother*, negotiations of "reality" in giving an account of one's experience are apparent, for instance in Xuela's witnessing of a

113 Xuela's European ancestry does not play a role here. Her father is a half-Scot and has very light skin and gray eyes, which Xuela sees as a physical marker of his cruelty – both are inherited from his colonialist father (*ABM* 186).

boy who vanished on their way to school: her schoolmate was lured into the river by the apparition of a beautiful woman, whom Xuela perceived not to be human, but "something that took the shape of a woman" (*ABM* 37), i. e., as a being from a more mythical realm. Rhonda Cobham recognizes the female figure as "the West African Mammywata, known in Jamaica as 'River Mumma' and associated loosely with such traditional African goddesses as the Ibo, Uhamiri/Idemili, and the Yoruba, Osun" (871). While Xuela insists that "I believed in that apparition then and I believe in it now" (38), the absence of further such mythical elements in the novel is conspicuous. One could read this absence as performing the success of Western policing of discourse. However, Cobham explains that in Caribbean literature, folk culture has a difficult status as it has an ambivalent relationship not only with European tradition that trivializes Caribbean practices, but also with African sources. Of the water deity Cobham writes, "Mammywata as a symbol of Africa may signify the lost homeland, a source of cultural resistance and material well-being, or a way of having one's self-worth affirmed. But Africa, like Mammywata is also a place that forgot to protect her own; that devoured her children in the pursuit of the promise of wealth" (875). Ultimately, in *The Autobiography of My Mother*, all of this intersects in who determines "reality." Dissuaded from talking about her witnessing of Mammywata's abduction of the schoolboy, Xuela realizes that "reality" here is defined by colonial holders of power. As Xuela puts it, "[e]verything about us is held in doubt and we the defeated define all that is unreal, all that is not human [...]. Our experience cannot be interpreted by us, we do not know the truth of it" (*ABM* 37). While the example here is one of myth, the doubting of "reality" is extended to all aspects of life.[114] Thus, every account that the "defeated" could possibly give of themselves is already discredited before it is even stated. This makes self-narration a project contesting the power of who polices discourse, i. e., the power of the colonizer.

The colonial policing of discourse also takes part in the definition of form. *The Autobiography of My Mother* explicitly comments on the dictations of genre in an episode in Xuela's school life: at the age of seven, Xuela is taught the rules of one specific genre – "the principles involved in writing an ordinary letter" (*ABM* 18). Its five aspects of addresses, date, salutation, main body, and closing solely focus on the formal features of the genre. Xuela comments on the futility of such a subject taught to her, as it was common knowledge that she – in her abject position – would not need to employ this genre in her life. Exactly this futility elicits an al-

114 Bart Moore-Gilbert points out that it is "a consequence of its political history [that] Caribbean subjectivity is [often] presented as caught on several levels between the 'real' and the 'fictive,' domains which cannot, in any case, be easily distinguished themselves" ("A Concern" 99–100).

most perverse "satisfaction" in her teachers (*ABM* 18–19), which again aligns the interest in genre with the figure of the colonizer and correlates the rules of genre with a colonizing force. Moore-Gilbert notes that literary scholars have indeed considered autobiography to be a thoroughly "Western" genre in its supposedly defining conventions of the category, such as the "unity and sovereignty of the autobiographical subject" ("A Concern" 93) that establish a "'value system' which underpinned and encouraged colonialism, notably its demarcation of the 'properly' human and appropriate means to narrate that privileged identity" (Moore-Gilbert, "A Concern" 92).[115] Jana Evans Braziel even links the sheer urge to categorize texts into genres to Western Enlightenment ideals which in other fields of study classified plants, animals, and even established the theoretical basis for the misguided notion of different human races (*Caribbean Genesis* 4–5). I do not imply to read performances of genre conventions in *The Autobiography of My Mother* in opposition to "Western" autobiography. Nor do I regard "Western autobiography" as a monolithic entity, as Moore-Gilbert's phrasing of "unity and sovereignty" might suggest. Such a strictly categorized and demarcated concept of autobiography nonetheless offers itself for consideration as Kincaid's own text, *The Autobiography of My Mother,* engages with a concept of generic rigidity that is arguably imposed by and aligned with colonial structures in order to transgress and transpose such boundaries.

In spite of restrictions of expression, Xuela masters the speaking of herself. She starts by appropriating the genre of letter writing at school. Instead of copying out letters of others (presumably English literary paragons deemed fit by her teachers), Xuela composes her own letters to her father who remained absent from the early years of her life. She describes her misery of being an outsider at school and at her foster home and wishes to be reunited with a loving father. Beginning her letters with "My dear Papa" (*ABM* 19), Xuela writes of herself within the tradition of nineteenth-century English orphan literature about impoverished children who will eventually be restored from hardship, as for instance exemplified by orphaned characters throughout Charles Dickens's works. While Xuela refuses to mindlessly copy and thus repeat the written works of English "role models," she nonetheless takes part in established forms of writing – here that of the orphan's letter. With this, *The Autobiography of My Mother* points to the ubiquity

---

115 Moore-Gilbert here specifically refers to Georges Gusdorf, Roy Pascal, and Richard Coe, all three of whom posit autobiography as essentially European (see Moore-Gilbert, "A Concern" 92 and 105; see also Moore-Gilbert's 2006 article "Western Autobiography and Colonial Discourse: An Overview"). Smith and Watson also maintain that autobiography "is a term for a particular practice of life narrative that emerged in the Enlightenment and thus became canonical in the West" (*Reading Autobiography* 3).

of writerly traditions from the colonial "mother country," but their hegemony is decentered when the Afro-Caribbean child inscribes herself into a literary history that generally works to marginalize her, when she puts herself at the center of the orphan narrative. She points to her precarious situation as an abandoned child through text, a medium that so far could not serve her (as expressed in the perceived futility of letter-writing). Now centering herself in such a literary tradition, Xuela can utilize it. She does not write these letters in all seriousness, as she does not expect them to reach their addressee. Rather, she writes them to express herself. The girl hides her letters under a stone by the school gate, where they are found by a spiteful classmate, who turns them over to the teacher. Xuela's fellow student had meant to hurt her as he expected her to be punished for writing about her maltreatment at school. The exposed letters, however, indeed make their way to Xuela's father, who promptly arrives to collect the forgotten daughter with promises of a new life. As the orphan story prescribes, Xuela is saved from her current plight. She recognizes that her act of writing of herself elicited the change in her life: "[H]owever unconsciously, I had, through the use of some words, changed my situation; I had perhaps saved my life. To speak of my own situation, to myself and others, is something I would always do thereafter" (*ABM* 22).

In *The Autobiography of My Mother*, Xuela narrates her life – foremost her relationships with her parents and with her lovers – from her own perspective. Moreover, the novel hardly contains any dialogue. Xuela does not grant much space to any voice but her own. The pronoun "I" certainly is one of the most used words throughout the book. However, (fictional) autobiography is not all that is at stake here. While the narrating voice may be identified as that of the fictional character Xuela, an uncanny entanglement of her life story with those of people outside her diegetic world remains, which multiplies the references of Xuela's narrating "I."[116] As Alison Donnell rightly points out, any avid reader of Kincaid's works, who knows the author's biography as well as her other fictional and nonfictional works, might easily conclude that *The Autobiography of My Mother* depicts the life of Kincaid's own mother, Annie Drew, née Richardson (127). However, it is just as obvious that Xuela could not possibly be a faithful rendition of Kincaid's mother, since the protagonist of *The Autobiography of My Mother* refuses to bear children and never left her family to move from Dominica to Antigua as Mrs. Annie Drew did. Rather, some basic information from Kincaid's family history

---

116 The confluence of fact and fiction is also furthered by the inclusion of a photographic portrait on the book's cover. Through its placement it suggests itself as a depiction of the protagonist, which of course is impossible as Xuela is a product of fiction. See chapter 3.3 for an analysis of the play with authenticity through the relationship of image and text in *The Autobiography of My Mother*.

is taken up by the novel, such as the structures of family relationships (for instance the difficult father-daughter-relationship between Annie Drew and her father), Annie Drew's childhood in Dominica, or the family names (e.g., Richardson) that recur throughout Kincaid's works.[117] This certainly suggests a relationship between Kincaid's family history, her personal biography, and her texts' characters, but it does not make them autobiographical. Nonetheless, suggestions of referential entanglements of the fictional character Xuela with actual people (Annie Drew and Jamaica Kincaid) add to the positions from which the life of Xuela is narrated *within* the "I" of *The Autobiography of My Mother.*

Sidonie Smith and Julia Watson posit that narrators in general are "a composite of speaking voices, the 'I' is a sign of multiple voices" (*Reading Autobiography* 60). Taking up Mikhail Bakhtin's term "heteroglossia," Smith and Watson assume that any autobiographical "I" consists of more than one voice.

> Heteroglossia, once incorporated into the novel (whatever the forms for its incorporation), is *another's speech in another's language,* serving to express authorial intentions but in a refracted way. Such speech constitutes a special type of *double-voiced discourse.* (Bakhtin, *Dialogic Imagination* 324, emphasis in original)

The question of who is speaking and who is being spoken about also stays relevant when leaving aside any speculations regarding autobiographical aspects of Kincaid's novels. In *The Autobiography of My Mother,* the "I" appears to contain more than one voice. As noted above, this multiplication of voices already starts in the book's title, which suggests two first-person speakers – a daughter within the possessive pronoun "my" and a mother within the defining morpheme "auto" of "autobiography." This tension between mother and daughter figure is continued in the first sentence of the novel. The text only partly adheres to the traditional notion of autobiography in beginning with its protagonist's birth. In fact, the first sentence of this text speaks to the mother: "My mother died at the moment I was born," are the first words of *The Autobiography of My Mother* (3). Considered in combination with the title, it remains unclear whose life narrative will unfold in the course of the book, as here the first-person narrator initially gestures towards a mother, which suggests that it might be her life that is to be told. However, it soon turns out that the "I" foremost relates to Xuela, who is thus aligned with the "auto" of the book's title. In this interpretation, the title "The Autobiography of My Mother" would be spoken from the position of a daughter, albeit one who does not exist within this diegetic world, as Xuela never bore children in

117 For instance, the names Annie, Richardson, or Potter occur repeatedly in variations in *Annie John, Lucy, The Autobiography of My Mother, My Brother,* and *Mr. Potter.*

this narrative of her life. If Xuela now speaks to her dead mother in the first sentence of the narrative, the text indeed addresses three generations of women: Xuela herself, her mother, and her (nonexistent) daughter(s). In this way, the text repeats the heteroglossic confusion that autobiographical speculation might evoke by multiplying references not only gesturing towards an extradiegetic world, but also within the diegesis.

A generating of generations through a heteroglossic narrative voice continues throughout *The Autobiography of My Mother.* For instance, the occasional departure from self-narration to the direct addressing of a "you" points to a multiplication of narrative voices: while the first-person pronoun takes on a distinct presence throughout the novel for its sheer frequency, at times it is interrupted by the emergence of a "you." As is to be expected of most texts, the second-person pronoun occurs regularly in this narrative, typically in direct speech. Another common instance of the pronoun being used is the reference to an unspecified person, i.e., pragmatically it is used in the form of an in impersonal pronoun, meaning "one." I read such uses of the pronoun "you" as addressing a general circumstance in life, albeit from Xuela's perspective.[118]

Something else is at play when "you" occurs in prominent clusters. Several times throughout the novel the second-person pronoun is used in a conspicuously frequent manner within a few sentences (for instance *ABM* 59–60, 69, 79, 202). Here, the possible addressees are multiplied within the pronoun "you," just as speaking positions are doubled and tripled in other passages. In these cases, the text is either preoccupied with the importance of names or with the mother figure (or with both), as in the following example:

> And *your* own name, whatever it might be, eventually was not the gateway to who *you* really were, and *you* could not ever say to *yourself,* "My name is Xuela Claudette Desvarieux." This was my mother's name, but I cannot say it was her real name, for in a life like hers, as in mine, what is a real name? My own name is her own name, Xuela Claudette, and in the place of Desvarieux is Richardson, which is my father's name; but who are these people Claudette, Desvarieux, and Richardson? To look into it, to look at it, could only fill *you* with despair; the humiliation could only make *you* intoxicated with self-hatred. (*ABM* 79, emphasis mine)

118 For instance, when relating the life and desires of Madame LaBatte, Xuela's host and employer when the young woman attends school away from home, a woman who so desperately wishes for a child that she wants Xuela to bear her own husband's, Xuela notes, "[t]o want what you will never have and to know too late that you will never have it is a life overwhelmed with sadness" (*ABM* 76).

The general statement about one's own name at the beginning of the quote gradually turns into a direct address of Xuela the mother, who is also named Xuela, as the sentence nears an insertion of imagined direct speech, which in the first person could hypothetically be uttered by Xuela the mother, as the use of her full name signifies. The following sentences return to the narrating voice of Xuela the daughter, also set in the first person, musing about the generational identity of names. The last sentence of the passage again addresses a "you," and it remains unclear whom it refers to. It might either be a generic you, a direct address of Xuela the mother, or the expression of Xuela the daughter's sentiments in the second person. The back and forth between pronouns entangles possible references and thus embodies both the pronouns "I" and "you" with different voices and addressees. Such a multiplication is already anticipated in the doubling of names, which points to a shared personal history in the first name and to a shared collective history of colonial subjugation, as the surnames Desvarieux and Richardson are assigned to both Xuelas through colonial power structures.[119]

In other cases, the shift of the narrative voice into the second person not only points to yet another position from which the narrative of Xuela is told. For instance, in the following quote, the second-person pronoun refers to Xuela Richardson: "It was that time of day when all you have lost is heaviest on your mind: your mother, if you have lost her; your home, if you have lost it; the voices of people who might have loved you or who you only wish had loved you; the places in which something good, something you cannot forget, happened to you" (*ABM* 69). Although all conditions here are circumscribed by an "if," they directly apply to Xuela's experience. In this, the second-person pronoun directly addresses the protagonist of *The Autobiography of My Mother*. Thus, these passages could be read as not being uttered from Xuela's own position, but from an outside perspective speaking towards Xuela, which suggests a temporary departure from Xuela as the narrative instance. At the same time, in terms of a generic you, this passage may as well be read as speaking to common experience of human beings, which points to a simultaneity of a collective and an individual within the "you." Ultimately, not only is the narrating voice heteroglossic, but those who they are speaking to and about are equally multiplied.

---

119 Richardson was the name of Xuela's (the daughter's) paternal Scottish grandfather who abandoned his Dominican family after stealing their hard-earned savings (*ABM* 194–195). Claudette and Desvarieux are names that Xuela the mother was given by a nun "who was on her way to wreak more havoc in the lives of the remnants of a vanishing people," as Xuela the narrator puts it (*ABM* 80), pointing to the colonizing project of Christian missionary work. This nun found Xuela the mother as a baby abandoned outside her convent's doors and decided to name the girl after herself.

While *The Autobiography of My Mother* ostensibly narrates the life story of Xuela, I argue that the text really is concerned with the narratives of women *like* Xuela. Xuela Richardson here stands in as a "literary witness" for a collective of women who have the same history and thus live under the same social conditions. With the figure of the "literary witness," Leigh Gilmore proposes to expand the testimonial archive beyond the actual texts of testimony to include literary texts. This, she argues, "expand[s] the limits of what it means to acknowledge and grieve the losses of history, and offer a traumatic witness capable not only of injury but also of speech, [which makes it] possible to see them in the articulation of what is always on the verge of disappearing: the human subject of historical and intimate trauma" ("What Was I" 83). Gilmore is interested in such literary witnesses as Xuela, who is her case study in her article, for their broadening of the archive considered in humanitarian discourse. In the context of my inquiry into genre, Gilmore's concept of the witness is elucidating, as it points to Xuela's capacity to speak for others. As Gilmore writes, "[t]he power of the first-person witness [...] rests on both the singularity and the wider representative capacity of the witness. In speaking to and for many, first-person accounts expand human rights beyond the frame of the individual" ("What Was I" 77). Gilmore notes that Xuela's statement that "all that is impersonal I have made personal" (*ABM* 228) expresses the narrator's project to voice a collective, "to particularize violence, to give it not only a human face and form but also a voice, a record made for those who cannot offer an account" (Gilmore, "What Was I" 80). Gilmore's reading of Xuela as a literary witness then arrives at the same conclusion that *The Autobiography of My Mother* multiplies voices within Xuela's, as my own analysis, which comes to the same conclusion by focusing on the literary form and techniques of the novel, particularly on heteroglossia and the redirection of references.

The intermingling of personal and collective becomes most obvious at the end of the narrative, when Xuela directly references her speaking as a representative for many:

> This account of my life has been an account of my mother's life as much as it has been an account of mine, and even so, again it is an account of the life of the children I did not have, as it is their account of me. In me is the voice I never heard, the face I never saw, the being I came from. In me are the voices that should have come out of me, the faces I never allowed to form, the eyes I never allowed to see me. This account is an account of the person who was never allowed to be and an account of the person I did not allow myself to become. (*ABM* 227–228)

Xuela merges the lives of three generations in a textual motion from herself to her mother and her nonexistent own children. It also includes the possible life trajectories she did not take. Xuela's autobiographical gesture to give an account of her life thus goes beyond the singularity of one life and she speaks to those of many. A collective is cospeaking with the voice of Xuela, as well as Xuela tells herself through others. Xuela becomes a container and mouthpiece for the voices of others who cannot speak (for) themselves as they could not come into being or were obliterated by history. Xuela thus expresses possibilities of the past, present, and future, indicating their continuities. She speaks not only *to* others ("you"), but also *with* and *for* them. The singularity of Xuela in this does not dissipate. Making the collective personal rather pluralizes Xuela's singularity when she contains and expresses a multitude of voices.

In conclusion, in this text speakers and addressees are multiplied within the pronouns "I" and "you," as well as they blend with each other. The identities of "mother" and "daughter" are intermingled within the diegetic world of *The Autobiography of My Mother*, but it is also suggested that the author Jamaica Kincaid and her own mother Annie Drew are part of the referential multitude of "I" and "you." Ultimately, however, my concern has not been not to disentangle this net of references. Rather, its existence itself, created through play with expectations and conventions of life writing, points to life narratives between *autos* and *bios*. They expand beyond the singularity of one life and speak to and for many. Xuela's telling of her self with and through others opens up the text from generic fixation to the different possibilities of speaking of one's own reality. *The Autobiography of My Mother* pretends to pretend to be an autobiography. While it seems to offer itself as an example of the genre, at the same time it clearly does not, and thus highlights the inability of categorizing and classificatory systems to transport (post)colonial experience that is simultaneously individual and shared by a collective. *The Autobiography of My Mother* performs a break with "Western" tradition, and Xuela provides one possible answer to the question posed by Smith and Watson: "If this autobiographical 'I' is a Western 'I,' an 'I' of the colonizer, then what happens when the colonized subject takes up a generic practice forged in the West and complicit in the West's romance with individualism?" ("Introduction" 28). At first, Xuela's narrative assumes the limits and rigidity of autobiography as a fixed (and fixing) genre. Such fixity resonates with the general colonial project of classifying and categorizing – from plants to people. The novel's performed turn against the rules of genre represents a rebuttal of colonial ideology. *The Autobiography of My Mother* thus exemplifies Kamau Brathwaite's expression that "the hurricane does not roar in pentameters" (10), speaking to the need for different literary forms to express (post)colonial Caribbean experience. The very contamination and confusion of genre and reference and the multiplications of "I" and "you," however, in their performance of breaking with colonial epistemological categories, open the text up to possibilities of self-determined representation and narration.

A Yellow Dress and a Brown Girl: Writing Between Subject and Object

"Biography of a Dress" is a short text by Kincaid that offers itself for an inquiry into genre with its very title.[120] Its evocation of life writing centering on a lifeless object raises questions such as "How might a biography of an object be written?" or "Is a biography not usually written about human beings?" Thus, the title confounds common genre expectations. The text is written in the first person and relates the events leading up to its narrator's second birthday, on which she is given a dress to wear which was made specifically for the purpose of celebrating the anniversary and photographing her in it to commemorate the day. A photograph of a toddler in a dress is included with the text, suggesting that this indeed is the image referred to by the text. This relationship of first-person narrator, relating memories of herself, in combination with a photograph which supposedly is one of the narrator and thus establishes her as a real person, evokes the genre of autobiography.[121] This runs counter to the text's title and questions who really is at the center of the text – the dress, as proclaimed by the title, or the first-person narrator, who ostensibly positions herself as central within the text.

The attempt to determine a generic category for "Biography of a Dress" is not aided by its publication history: the eight-page text was printed in a 1992 issue of the literary magazine *Grand Street*. The publication in this context suggests an intended audience of cosmopolitan intellectuals, interested in national and international literary and visual arts. The placement of "Biography of a Dress" in this magazine does not illuminate questions of genre, but rather leaves them open to be performed by the piece itself.[122] The text begins on a right-hand page of the

---

120 From here on, I use "BD" to refer to Kincaid's "Biography of a Dress."

121 Jana Evans Braziel determines that, "[i]n the short story 'Biography of a Dress,' [...] Kincaid poses the question, what is biography?" (*Caribbean Genesis* 19). More than that, I argue, the text signifies on the relationship of subject and object in the complex terrain of coloniality through a negotiation of genre and its perceived boundaries.

122 A contextualization with the other texts in this issue does not help much, either. The issue was published (1992) before the magazine started to put out thematic issues (Winter 1994), so that the contents are brought together without overt correlation: "Biography of a Dress" is wedged between two poems, "The Tale of Murasaki" by Martine Bellen, and "Derision" by Horrah Pornoff. What is noteworthy, however, is that while *Grand Street* is interested in both written and visual arts, Kincaid's work here apparently was not chosen for publication for its photograph, but rather for the text: other visual artwork in the issue is printed on especially glossy pages of thicker paper, which are reproductions of higher quality than Kincaid's photograph, simply printed on the regular paper used for all pages containing text. All this leads to the conclusion that in the process of publishing "Biography of a Dress" the text was regarded as of higher interest than the image. The image is an accessory to the text. In my analysis, however, the photograph is of central importance.

magazine spread. The author's name, "Jamaica Kincaid," is placed at the top of the page, followed by the text's title "Biography of a Dress." Opposite, on the left-hand page, a photograph is printed on a gray background which covers the whole page and frames the photograph on the book page as a passe-partout would in a picture frame. The black-and-white photograph shows a small girl standing on top of a small desk in front of a blurry painted background with a white, trompe-l'oeil balustrade in the bottom right that gradually vanishes as the eye moves left. If familiar with Jamaica Kincaid, the reader would quickly detect similar facial characteristics between the girl depicted and public photographs of the author, which easily implies that the photograph might indeed portray Jamaica Kincaid as a young girl. *Grand Street*'s list of illustrations at the very end of the issue details the photograph as "Jamaica Kincaid, c. 1951. Courtesy of Jamaica Kincaid" (216). Far removed from the print of the photograph itself, however, this note does not immediately affect how it is first read on page ninety-two of the book-length publication. As with the photographs in *Mr. Potter* and *The Autobiography of My Mother*, discussed in chapter 3.3, the placement of the photograph here, too, implies a connection between text and image through proximity. Yet even with the confirmation that the photograph shows the author as an infant, it would be imprudent to assume this connection to be reliable, as Kincaid's works generally toy with notions of authenticity and fiction through a play with conventions.

The first lines of the text of "Biography of a Dress" strengthen the impression that text and photograph depict the same person: "The dress I am wearing in this black-and-white photograph, taken when I was two years old, was a yellow dress made of cotton poplin" (BD 93). This quote brings together all five elements mentioned here so far: the dress, the photograph, and the "I" of the text, which refers to both the person depicted in the photograph, the narrator of the text, and to the author Jamaica Kincaid. The concurrence of narrator and author in the first-person pronoun as well as the attribution of the "I" to the image suggest that the text at hand indeed is autobiographical. However, from the very first line of the text, the personal pronoun referring to a human being is set behind "the dress," which again puts the dress into a premier position, suggesting an importance of the dress in opposition to the person wearing it, just as the text's title does.

The narrative situation of the text initially appears as one typically occurring in autobiography.[123] Situated in a present time, a narrating "I" relates the ex-

---

For its prominent placement and its reciprocal relationship with the text, I regard it as equally essential to the whole of the work as the text.

123 The text's effects here seem to work, for instance J. Brooks Bouson reads "Biography of a Dress" as an "autobiographical story" (43). Bouson unquestioningly equates the protagonist and the narrator of "Biography of a Dress" with the real person and author Jamaica Kincaid. Bouson

periences of a narrated "I," which occurred in the past.[124] The narrating "I" is forty-three years old (BD 97) and claims to remember the photograph being taken, as well as the preparations leading up to this event, when she was two years old. The temporal distance is showcased by parentheses and inserted commentary. Throughout the text the narrating "I" comments on the events and circumstances connected to the portrait of her two-year-old self in brackets, as for instance in the following passage:

> But I was then (not so now) extremely particular about what I would eat, not knowing then (but I do now) of shortages and abundance, having no consciousness of the idea of rich and poor (but I know now that we were poor then), and would only eat boiled beef (which I required my mother to chew for me first and after she had made it soft, remove it from her mouth and place it in mine), certain kinds of boiled fish (doctor or angel), hard-boiled eggs (from hens, not ducks), poached calf's liver and the milk from cows, and so would not even look at the boiled cornmeal (porridge or fongie). (BD 93–94)

This sentence from the first two pages of "Biography of a Dress" contains six parentheses. Initially, they signal a temporal remove of the narrating "I" from the narrated "I," clearly separating the two when the main text refers to a "then," while the commenting parentheses locate the narrating instance in a "now."[125] Moreover, when the narrating "I" comments on her own narration, evoking a temporal distance and layers to her story, the narrating "I" locates herself in two different spaces within the text. While the narrator stays the same, she appears to be split up into two voices.

Julia Emberley notes about the parentheses that "[h]istorical memory, both personal and impersonal, necessitates a break, a rupture. In the discontinuous present Kincaid brackets off the painful memory" of the experience of having the photograph taken of herself (469). I argue that it is not the memory which is bracketed off from the present, but rather that the present is inscribed into the narrative of the past. The "I" in the main text relates the story of its two-year old protagonist. The other "I" within the parentheses comments on this narrative and thus inserts her contemporary views and knowledge. In this, the narrating "I" tells of herself in

---

even goes so far as reading "Biography" as relating the young life of Annie John (43), protagonist of Kincaid's novel *Annie John*. In this, Bouson regards these two textual figures as the same and as converging in Jamaica Kincaid.

124 In using the terms "narrating and narrated 'I,'" I am here referring to the categories laid out, for instance, by Sidonie Smith and Julia Watson in *Reading Autobiography*, esp. 58–64.

125 In this juxtaposition of temporal levels, "Biography of a Dress" operates much in the same ways as Kincaid's later novel *See Now Then* (2013). See chapter 1 on "Time and Temporalities in *Mr. Potter* and *See Now Then*."

the present. It thus emerges as a distinct presence which I regard as a second narrated "I." In the parentheses, the narrating "I" finds the space to make herself known.

Yet the older narrating "I" does not appear to be a reliable instance, when an excess of details calls the accuracy of her memories into question. It seems rather unlikely that a forty-three-year-old narrator would be able to recall the experiences leading up to her second birthday so specifically, such as the interactions with a store clerk who sold the material for her birthday dress (BD 94), or the specifics of having her photograph taken at a studio (BD 99–100). While the experience of pain, such as having her earlobes pierced with two hot thorns (BD 97), may certainly be traumatizing and thus impress a lasting memory on such a young person, it nonetheless appears unlikely that the details of the events before and after would be as present after four decades, especially considering that the narrator repeatedly points out how her birthday was marked by triviality (for instance BD 94, 97, or 99). The impression of exaggeration bordering on fictionality lets the narrator appear rather unreliable, which begs the question of constructedness and – even more so – the purpose of producing not only the narrative of the two-year-old girl, but also the memory work of the older narrating "I." It remains unclear from which position the narrated "I"s are constructed, which suggests yet another narrating instance within the first-person pronoun.

Another indication for the text's fictionality is the magazine's note on the origin of the photograph as "Jamaica Kincaid, c. 1951. Courtesy of Jamaica Kincaid" (216). As mentioned, the similarity of the girl depicted with the public photographs of the author Jamaica Kincaid already suggests the identity of the two, which is supported by the denotation that the image shows "Jamaica Kincaid" (although "Elaine Potter Richardson" might have been more accurate, as the picture would have been taken long before Kincaid's renaming). Richardson/Kincaid was born on May 25, 1949. The two-year difference between 1949 and 1951 once more suggests the depiction of the author as a two-year-old. The "c.", abbreviation for circa, however, calls aspects detailed in "Biography of a Dress" into question: if this really was the photograph taken on Kincaid's second birthday, the date of its origin could be provided more specifically than only the year. Given the text's insistence on the particular date of the photographing, the image source, which includes an oddly vague "circa," would be outright absurd. The "c." thus suggests a certain extent of fictionality to "Biography of a Dress," which in turn suggests that the narrating "I" possibly might not be who she claims. Yet another narrating "I" thus enters the narrative, one who constructed the narrative as that of a forty-three-year-old "I" who, through a photograph, is linked to a younger narrated "I." While this multitude of "I"s is a key element of any autobiographical text, "Biography of a Dress" conspicuously stages them through parentheses, and thus the narrating

"I" is not simply a vehicle to produce the story of the narrated "I," but it overtly presents the heteroglossic multiplicity of voices within the "I."

The separation of narrating and narrated "I" is later described by the text as having originated in an experience of immense pain. To commemorate her daughter's second birthday, her mother had the girl's ears pierced on that day. On the one hand, the pain here is physical, but on the other, it is also that of feeling violated, as the narrator muses whether her mother may inadvertently have "meant to express hostility and aggression toward [her]" (BD 97). Suffering this pain then has a dissociative effect on the girl: "[I]t was then that was the first time that I separated myself from myself, and I became two people (two small children then, I was two years old), one having the experience, the other observing the one having the experience" (BD 97). The separation the narrator describes is a method of self-preservation. The split between the physical experience and the cognitive understanding of this pain, which is described as an out-of-body experience, suggests that the entity relating the experience in text may be associated with the so-called observer who is distanced in and by the heteroglossic parenthetical comments. Considering this split within the "I," I reason that the parenthetical writing style of "Biography of a Dress" with its plurality of narrated and narrating "I"s mirrors the psychological state of the girl as well of the narrator.[126]

Such dissociation from oneself is a central aspect of colonial experience, as theorized for instance by W. E. B Du Bois in his concept of "double-consciousness." In *The Souls of Black Folk*, Du Bois details his realization that he was Black when he was a schoolboy, which came about in a situation of exclusion and ostracization. Du Bois felt shut out from the world of his white school mates by a "vast veil" (2). Later, Du Bois explains his perceived position as doubled:

> [T]he Negro is [...] born with a veil, and gifted with a second-sight in this American world, – a world which yields him no true self-consciousness, but only lets him see himself through the revelation of the other world. It is a peculiar sensation, this double-consciousness, this sense of always looking at one's self through the eyes of the others, of measuring one's soul by the tape of a world that looks on in amused contempt and pity. One ever feels his two-ness, – an

---

126 Emberley even muses that Kincaid entitled the text "'biography' of a dress and not the autobiographical account of her second birthday" (469) with the intention to separate herself from a memory of violence, symbolized by the dress. While this thought is an intriguing one, I argue that the genre contextures are much more intricate. As I show in this analysis of "Biography of a Dress," Kincaid does not simply attempt to distance herself from the object of the dress (as Emberley puts it, 469), but her narrating self performs the distance between experiencing "I" and her cognition, as well as the distance between herself as a speaking subject and her former objectified self.

> American, a Negro; two souls, two thoughts, two unreconciled strivings; two warring ideals in one dark body, whose dogged strength alone keeps it from being torn asunder. (2)

In Du Bois's notion of double-consciousness, the internal split of the minority, excluded from the dominant group, is caused by the internalized white gaze. Being looked *at* as "the Other" leads to the separation from one's own consciousness by simultaneously looking at oneself through someone else's eyes.

Frantz Fanon later, in *Black Skin, White Masks*, takes up Du Bois's notion in the famous train episode in which he is framed and frozen by the white gaze of a boy as a "Negro." Fanon sees no other option but to distance himself from his body, which is predetermined by the onlooking white people in racist concepts of Blackness. Fanon feels forced to escape cognitively. He writes, that "[d]isoriented, incapable of confronting the Other, the white man, who had no scruples about imprisoning me, I transported myself on that particular day far, very far, from my self, and gave myself up as an *object*" (92, emphasis mine). In Fanon's explication, the inner split not only leads to "two-ness," as Du Bois put it, but to the objectification of the self – the white gaze leads to becoming less than human, i. e., to becoming a thing.

The separation of the girl from herself in "Biography of a Dress" is a protective measure, but it is one that is not innocuous, as it filters and thereby distorts any experience. Moreover, it affects the constellation of narrating and narrated selves:

> And the observer, perhaps because it was an act of my own will (strong then, stronger now), my first and only real self-invention, is the one of the two I most rely on, the one of the two whose voice I believe to be the true voice; and of course it is the observer who cannot be relied on as the final truth to be believed, for the observer has woven between myself and the person who is having an experience a protective membrane, which allows me to see but only feel as much as I can handle at any given moment. (BD 97–98)

The observing self here is described as yet *another* instance separated from the "I" in this quote. The speaking "I" positions itself as central, separated from the experiencing self as well as from the cognitive self of the observer, but nevertheless as the location in which experiencing "I" and observing "I" converge. Simultaneously, this "I" is also an experiencing "I" as she realizes her nonexperience, while also separated from the generally experiencing self. As such, this "I" is located outside experience and observation, but by having awareness of both, she also encompasses them both. In an economy of the ability to speak, the "center" and the "observer" relate the life of the experiencing "I," as well as their own thoughts on it and on their present selves, while the two-year old experiencing self remains quiet. The little girl is the nonspeaking object of interest of the speaking subjects. Fittingly, Smith and Watson posit that the "narrated 'I' is the object 'I'" (60).

Considering the autobiographical characteristics of "Biography of a Dress," the text's narrating selves may be located in the author's name "Jamaica Kincaid," with which the text is overwritten. According to Philippe Lejeune, the "first person" is defined by "reference" and by "utterance:" "The 'I' refers, each time, to the person who is speaking and whom we identify *by the very fact* that he is speaking,"[127] Lejeune notes about the referentiality of the "first person" (9, emphasis in original). "In printed texts, responsibility for all enunciation is assumed by a person who is in the habit of placing his *name* on the cover of the book" (Lejeune 11, emphasis in original). Hence, the utterance of "I" is by convention assumed to be that of the person identified by name. In the "*initial section* of the text [...] the narrator enters into a contract vis-à-vis the reader by acting as if he were the author, in such a way that the reader has no doubt that the 'I' refers to the name shown on the cover, even though the name is not repeated in the text" (Lejeune 14, emphasis in original).

The connection of narrator, protagonist, and author through reference and utterance of the first-person pronoun is then supported by the connection with the photograph. The first line of the text ("The dress I am wearing in this black-and-white photograph") explicitly links the pronoun "I" and the girl in the photograph by way of the deictic determiner "this," the referential gesture of which is supported by the immediate proximity of the printed photograph on the opposite page, as well as by the correspondence of telling and showing in the presence of a dress in this photograph. This creates a referential relationship between depicted person, narrating I, and narrated two-year-old I. Here the selves now converge not only in the author name, as assumed by Lejeune, but also in the photograph of the child. The conspicuous resemblance between press photographs of the author and the image of the child also further the assumption of identity of the "I"s of the text and the depicted person in the photograph.

Still, the title turns the attention back to the dress, which again raises the question of the relationship of human and thing, subject and object. The text begins with the words, "[t]he dress," and goes on to describe this dress before attending to the girl wearing it: it "was a yellow dress made of cotton poplin (a fabric with a slightly unsmooth texture first manufactured in the French town of Avignon and brought to England by the Huguenots)" (BD 93). Starting out with the history of the fabric, the text then continues to relate the history of this particular dress: its fabric was bought by the narrator's mother (BD 94), together with yellow thread for embroidery (BD 95). The mother then took days to sew and embroider the dress by

127 Both Lejeune and his translator Katherine Leary use the masculine pronoun in reference to all authors and speakers throughout.

hand.[128] The narrator details: "the bodice of the dress appeared simple, plain, and the detail and pattern can only be seen close up and in real life, not from far away and not in a photograph" (BD 96). Indeed, considering the image provided, it is entirely impossible to make out such detail. In the black-and-white photograph it is even impossible to tell whether this dress is yellow at all; it might as well be white.

Within the photograph itself, the tension between girl and dress is just as prevalent as in the text: in stark contrast to the generally darker tones of the photograph, the dress stands out. It is centered right in the middle of the image. Shining very light, almost white, it eclipses the face and body of the girl wearing it.[129] The dress comes to the fore, while the girl's skin almost merges with the colors of the background. Ghostlike, the dress and the light shoes appear to float in front of a background of gray tones. This visual effect of the blurry gray hues, which lets the girl fade, parallels the objectification of the girl by the text through literary strategies and by the mother, who turns her daughter into a colonial object by acting *on* her.

At the photographer's studio, the girl is placed standing "on a table that made me taller, because the scene in the background against which I was to be photographed, was so vast, it overwhelmed my two-year-old frame, making me seem a mere figurine, not a child at all" (BD 99). In her perception, the girl is transformed from a human into the decorative object of a "figurine" – a lifeless small statue modelled after the shape of a human being. Indeed, it is not only the smallness of the girl's body against the tapestry, but the act of being photographed itself, particularly without voluntary participation, that objectifies the girl. "To photograph people is to violate them, by seeing them as they never see

---

128 In addition to the play with temporalities through the use of parentheses, the juxtaposition of photograph and narrative also relates to different temporalities, which on the one hand is based on the different capacities of the media, but on the other hand also on the specifics of the story told. As Nicole Matos similarly observes, image and text refer to two different time frames. "While the photograph itself captures a single moment […] the story captures a weeks-long chain of preparations, culminating in the final hours before the picture is snapped" (Matos, "The Difference" 844).

129 Braziel notices that in Kincaid's works, "the colors yellow and white often symbolize decay and decadence and death. This color-coded textuality manifests a powerful critique of the political, cultural hegemony of whiteness by Kincaid" ("Daffodils" 91). In the "Biography of a Dress," yellow is certainly associated with revulsion. The color is mentioned several times, and each time it describes an object that elicits disgust, such as a disliked food (fongie, BD 93), the intestines of a dead man (BD 94), or pus flowing from a sore on the photographer's cheek, which is associated with the icing on the girl's birthday cake (BD 100). The yellow dress joins the ranks of such repulsive objects. Moreover, it also symbolizes the imposing hegemony of whiteness, or rather Englishness, as Braziel observes, in that it represents colonial ideals encroaching on the body of the girl.

themselves, by having knowledge of them they can never have; it turns people into objects that can symbolically be possessed," as Susan Sontag puts it (14).[130] The driving force here is the girl's mother, whose bathing, powdering, and dressing her daughter for the photograph illustrates an internalized desire to adhere to colonial beauty standards.

The mother acts through the scripts internalized by her own existence in a colonial system. She mimics the colonizer's gaze and perpetuates it by acting it out on her daughter's body. In this way the girl's mother becomes an agent of coloniality who navigates within a colonial system enforced by material culture that extends to every aspect of life, as the dressing of her daughter exemplifies. Practicing her embroidery, for instance, the mother adorns her daughter's dresses with "species of birds she had never seen (swan) and species of flowers she had never seen (tulip) and species of animals she had never seen (bear) in real life, only in a picture in a book" (BD 96). The three species mentioned here are in their combination typically associated with European imagery, which aligns with the distribution of European "culture" in colonized countries, here that of Great Britain in the Caribbean.[131]

The mother acts as a colonizing instance towards the girl when she tries to have her daughter look like a "proper" English girl. The contemporary narrator of "Biography of a Dress" locates the origin of this yearning in a soap advertisement:

> [A] girl whose skin was the color of cream in the process of spoiling, whose hair was the texture of silk and the color of flax, a girl whose eyes gleamed like blue jewels in a crown [...] – that my mother saw, a picture on an almanac advertising a particularly fine scented soap (a soap she could not afford to buy then but I can now), and this picture of this girl wearing a yellow dress with a smocking on the front bodice perhaps created in my mother the desire to have a daughter who looked like that or perhaps created the desire in my mother to make the daughter she already had to look like that. (BD 96–97)

130 See chapter 3.1 for a detailed discussion of objectification through photography, especially in *Lucy*.

131 Another example from Kincaid's works would be Lucy's ire at daffodils upon seeing them in person for the first time, as they provoke memories of being forced to learn the poem "I Wandered Lonely as a Cloud" by William Wordsworth by heart. Lucy's anger at the poem is caused by her disapproval of colonial education, centered on subjects far from the real world (see *Lucy* 30). Similarly, the decorative plate painted with an English landscape, meant to symbolize "heaven," which is revered by Ma Eunice in *The Autobiography of My Mother* (9), transports English ideals via a commodity. See chapter 3.1 for a discussion of colonial imagery here.

This quote reveals that the fuss about preparing the girl and having her photographed is not to celebrate her second birthday, but to recreate the image of a white girl in a yellow dress, as known from a soap advertisement.

What affects the mother are the connoted meanings of the iconic message of the advertisement. In his analysis of a French pasta advertisement in "Rhetoric of the Image," Roland Barthes expounds the connotations of the symbolic message of the image, which is joined with the linguistic message and the noncoded (denoted) iconic message ("Rhetoric" 36). The pasta advertisement consists of an open shopping bag with packets of pasta, sauce, and parmesan, and fresh onions, tomatoes, and mushrooms arranged within and next to it. These visual elements make up the denoted iconic message. To describe the symbolic iconic message of the advertisement, i. e., the aforementioned elements' connoted meaning, Barthes contrives the term "Italianicity" ("Rhetoric" 41). He writes that "Italianicity is not Italy, it is the condensed essence of everything that could be Italian, from spaghetti to painting" ("Rhetoric" 48), which the articles in the shopping bag are meant to convey through their combination, arrangement, and color scheme (*tricolore*). In combination of the connoted and denoted meanings, the advertisement appraises not only packets of pasta, but a "feeling" of "Italianicity" that the consumer might acquire together with the tangible pasta.

Following this logic of the connoted signifier of the advertisement, I maintain that what the soap advertisement described in Kincaid's "Biography of a Dress" connotes is proper English "girlness." Considering the details of the iconic message – including the whiteness of the depicted girl's skin or the silkiness of her flaxen hair – the message of the advertisement is not simply: "Buy this soap!," but rather: "This soap signifies the capacity to cleanse and whiten ethnic markers to approximate white girlhood in aesthetic, moral, and sexual terms. If you desire the ascriptions of a white girl's body – buy this soap!" A personal hygienic product and desirability of "white girlness" are thus entangled in the symbolic message of the soap advertisement. In the "Biography of a Dress," the mother bathes and powders the girl (BD 98) before putting her into a yellow dress with a smocking (BD 99) in her attempt to recreate the image of the soap advertisement's white "girlness." The mother thereby repeats the internalized white gaze of the colonizer and thus objectifies her two-year-old child by attempting to reproduce an idealized image of English "girlness" through the body of her daughter, as well as through the photograph taken of her.

Yet the mother's attempts to create a whitewashed daughter for herself must remain futile, as the narrating "I" explains: "My skin was not the color of cream in the process of spoiling, my hair was not the texture of silk and the color of flax, my eyes did not gleam like jewels on a crown" (BD 96). The girl could not possibly fulfill her mother's desire to look like the girl in the soap advertisement. She can only be “almost the same but not quite,” to use the phrase that Homi Bhabha employs to define colonial mimicry (127). Almost as a side effect of attempted “sameness,” objectification thus lays bare the mechanisms of mimicry. It unveils its double vision that reveals the colonizer (Bhabha 126).

The text of “Biography of a Dress,” I argue, counters the objectifying attempts of the girl that culminate in the photograph. The preparations leading up to it as well as the photograph being taken participate in framing the girl in a vision of the colonized other. The image itself fixes the girl in this projected “English girlness.” Initially, with its staged confusion of genres and title, the text repeats the marginalization of the girl. She is displaced by the focus on the dress, and framed as helpless by the text when it speaks for her. The experiencing “I” is not allowed to speak; another voice narrates her, writes her biography, which is also her own, but the narrated “I” remains detached from the “I” capable of acting. When the autodiegetic voice multiplies heteroglossically, it becomes increasingly unclear who is speaking. While such heteroglossia at first appears to be typical for autobiographical texts, in “Biography of a Dress” it is so overtly presented that it points to a fragmentation within the “I” that is distinctly (post)colonial. As such, the text’s use of typical genre features formally reverts to the underlying issue of human beings regarded as objects in a colonial system. It speaks through an “I” that is heteroglossic and fragmented, an “I” who was compelled to work itself through layers of objectifying colonial scripts (including what defines autobiography), producing an experience that the “I” now brings to the fore, in speaking about herself at all. In this way, the “I” is also a “she,” demonstrating the mechanisms of objectification and establishing herself as a self-determined subject.

3 Images: Visual Engagements with History

Before she became an established writer of both fiction and nonfiction, Kincaid had intended to become a photographer. Although she eventually turned to text as her preferred medium, visuality still features strongly throughout Kincaid's works. Looking and seeing are central occupations of most of her characters as early as in the short stories collected in *At the Bottom of the River* or in *A Small Place.* Kincaid's fictional characters engage with images and with photographs in particular, above all in *Annie John* and *Lucy.* Other texts feature actual images, such as *A Small Place*, "Biography of a Dress," *The Autobiography of My Mother*, *My Garden (Book):*, and *Mr. Potter.*

The images in Kincaid's works have intricate relationships with the texts with which they are placed (or, in the case of ekphrases, they are brought forth by text in the first place). In a 1994 interview, Kincaid comments on connections of text and photography: "I was in college and thought I would be a photographer, and I used to write out my photographs [...] – what I would take and how I would set them up. [...] I would write down what I thought the picture should feel like. And I would try to take a picture of what I had written down" (Kincaid in Ferguson, "A Lot of Memory" 163). Such connections of the literal meaning of photography – "writing light" – and actual writing culminate in Kincaid's use of ekphrases, particularly in her later works, *My Garden (Book):*, *Mr. Potter*, and *See Now Then.*

Throughout Kincaid's oeuvre, visual media engage with personal and collective history when visibility in imagery is part of the metaphorical visibility of Black bodies in history. bell hooks explains that not to have one's picture taken means "to stand outside of history" ("Photography" 49), since, as one could add with Susan Sontag, "[t]o photograph is to confer importance" (28). Moreover, like written histories, "visual culture can play its part in redefining culture as a constantly changing, permeable and forward-looking experience of transculture, rather than as a clearly definable inheritance from the past" (Mirzoeff 132). I regard the visual art in Kincaid's works as such agents of the making of history – as a heightened form of what Hayden White calls *historiophoty*, "the representation of history and our thought about it in visual images" ("Historiophoty" 1193). Demonstrating the openness of visual art to interpretation and (re)signification, Kincaid's works explore the capacities they possess to represent history, including different competing versions. Constantly questioning the nexus of visual representations and (past) reality, these texts reveal a flexibility in the relationship of signifiers and signifieds in both visual and textual images. And in uncovering these unexpected flexibilities, the engagements with images and with text-image rela-

 https://doi.org/10.1515/9783111027500-005

tions take up the back-and-forth of the poetics of impermanence that I have identified, in the present book, as characteristic for all of Kincaid's oeuvre.

The following chapters investigate three different ways in which visuality occurs and engages with history in Kincaid's works: chapter 3.1, "Making and Unmaking Representations of the Past," analyzes images on the diegetic level of Kincaid's texts. The focus here lies on Annie John's iconoclastic acts and Lucy's production of images to demonstrate how the characters reject being framed by others in stereotypical (colonial) discourses, refuse to recognize colonial iconography, and seize the potential that photography offers to projects of self-making. The following chapter 3.2, "A Photo Album of History," is concerned with Kincaid's ekphrastic technique, in particular in *My Garden (Book):*. I argue that the book employs ekphrases, i.e., verbal representations of visual representations, as textual photographs that question the possibilities of representation in general and, more specifically, of representation in historical colonial contexts. Chapter 3.3, "Photographic Product(ion)s of History," focuses on the concrete photographic portraits that feature in *The Autobiography of My Mother* and *Mr. Potter* to indicate the production of new versions of the past through the interplay of seemingly authentic photographs and fictional texts.

3.1 Making and Unmaking Representations: Intradiegetic Images in *Annie John* and *Lucy*

Jamaica Kincaid's early novels *Annie John* and *Lucy* are concerned with images and visuality on the diegetic level. The visual practices of the texts' protagonists engage with both personal past and collective colonial history. Annie John's iconoclastic engagement with family photographs and with images that reference colonial history decenters the significance of the images' referents and, with that, the significance of past events. Annie resists being passively framed in both photographs and social roles. She insists on her own agency to interpret images and to interpret the past. Similarly, Lucy refuses to be framed by either exoticizing colonialist imagery or by stereotypes of immigrant workers and domestic help. She then returns the gaze of these views in actively taking up the camera and, in a repetition of colonizing gestures, appropriating her environs in the United States by photographing them herself.

This chapter explores how the protagonists' interactions with visual media, particularly with photographs, address personal past, colonial history, and the relationships of both to the present. As a seemingly truthful medium, photography might be thought to provide uninhibited access to the past. However, as I will show, in Kincaid's works photographs are not conceived of as faithful renditions of the past that offer up unambiguous significations. On the contrary, constant interpretative work is required for meanings to come forth. Here, meanings are not inherent to images; rather, the texts demonstrate how meanings are fundamentally attributed to images by the observer. Accordingly, Annie and Lucy do not use images to gain knowledge about the past. Instead, by manipulating and producing images, they negotiate their own subject positions in the present. My interest in the texts' discussions of visuality and in the characters' engagement with images is thus twofold: first, to explore the conception of visual representations as put forth by Kincaid's texts, and second, to examine the texts' understanding of personal past and collective history as it is displayed by Annie's and Lucy's visual practices.

In his seminal work *Camera Lucida,* Roland Barthes famously claimed that "in Photography I can never deny that *the thing has been there.* There is a superimposition here: of reality and of the past" (76, emphasis in original). Barthes emphasizes that the thing or person photographed must have existed and "been there" for the photograph to have come into being. In this, Barthes is not foremost interested in the photograph's iconicity, which signifies meaning by similarity (Peirce 157), but he does directly pick up on Charles Sanders Peirce's stipulation that "due to the photographs having been produced under such circumstances that they were physically forced to correspond point by point to nature, [...] they belong to the second class of signs, those by physical connection [indices]" (156). Following this, for Barthes the photograph is consequently proof of a past reality. As such, a photograph would indeed provide a form of access to the past. Susan Sontag, in contrast, notes that every photograph is already "an interpretation of the world as paintings and drawings are" (7).[132] Accordingly, photographs may provide access to a mediated past, but not to an alleged reality. Rather in accord with Sontag, the images described in Kincaid's texts disrupt the Barthesian notion of the photograph as a gateway, in that they consistently deny any inherent meaning. The connection of the photographic signifier to a real-life object (or person), which Barthes also maintains when he writes that "[a] specific photograph, in effect, is never distinguished from its referent (from what it represents)" (*Camera Lucida* 5),[133] is disrupted in Kincaid's works. One sees this, for instance, in *Annie*

132 By contrast, in "Rhetoric of the Image," Barthes insists that "[m]an's interventions in the photograph (framing, distance, lighting, focus, speed) all effectively belong to the plane of connotation; it is as though in the beginning (even if Utopian) there were the brute photograph (frontal and clear) on which man would then lay out, with the aid of various techniques, the signs drawn from a cultural code" (44).

133 Barthes does, however, concede that in special cases, it is ultimately possible to distinguish between photograph and referent: "at least it [the photographic signifier] is not *immediately* or

John, when components of the photographic sign are perceived individually, ultimately producing new indexicalities and, with that, different meanings. In every act of perception the images' significations thus need to be decoded anew, as they might constantly change in the eyes of their (different) beholders.

In *Lucy*, the family photo albums kept by Mariah, the protagonist's employer and surrogate mother figure in the United States, perfectly illustrate processes of meaning-making when emotional attributions to the medium (the album and the photographs it contains) prove more important than what the images signify. We see this, for instance, when Lucy notes that the albums fulfill the specific function of presenting a certain narrative of Mariah's family, and that of her and her husband Lewis in particular. The albums start with the initial meeting of Mariah and Lewis "in Paris in the shadow of the Eiffel Tower or in London in the shadow of the Big Ben or somewhere foolish like that. [...] And there were pictures of them getting married [...] and of their children just born in hospitals, and birthday parties and trips to canyons and deserts and mountains, and all sorts of other events" (*Lucy* 79–80). Mariah's photo albums present what Susan Sontag calls a "portrait-chronicle" of a family, "a portable kit of images that bears witness to its connectedness" (8). Lucy realizes that Mariah ascribes emotional value to her arrangement of the photographs, which purports a coherent story of her family life. Yet Lucy also understands that this narrative is created by Mariah and did not occur naturally, as Mariah might like it to appear. On her discovery that the seemingly perfect husband Lewis is carrying on an affair with Mariah's best friend Dinah, Lucy notes: "But here was a picture that no one would ever take – a picture that would not end up in one of those books, but a significant picture all the same" (80). By arranging the images to her liking and by not including all possible significant events,[134] Mariah creates patterns for the narrative of her family history.[135] Mariah here uses the images symbolically: she foregrounds a cultural code of family photographs that signifies harmonious relationships by collecting them in photo albums, by placing others "all over the house" (*Lucy* 12) on display for the family and possible visitors, and further in the composition of the photographs, when the family members' "six yellow heads of various sizes were bunched as

---

generally distinguished from its referent [...]: it is not impossible to perceive the photographic signifier [...], but it requires a secondary action of knowledge and reflection" (*Camera Lucida* 5).

134 That said, Mariah could not have included photographs of Lewis's marital transgressions even if she wanted to, since first, such images were never taken, and second, Mariah is yet unaware of these events.

135 Mariah's structuring of the visual narrative here is highly reminiscent of Hayden White's notion of the emplotment of historical events as the transformation of sequences into narratives, as chosen by the historiographer (see White, *Metahistory* xii).

if they were a bouquet of flowers tied together by an unseen string. In the pictures, they smiled out at the world, giving the impression that they found everything in it unbearably wonderful" (*Lucy* 12). The family photographs thus "turn the past into a consumable object, [and] are short cut. Any collection of photographs is an exercise in Surrealist montage and Surrealist abbreviation of history" (Sontag 68). Lucy's recognition that by no means *all* major events of Mariah and Lewis's relationship are recorded in the albums indicates that the images' intended symbolic meanings originate with Mariah.

However, Mariah does not have an authorial prerogative of interpretation, as the meanings or signifieds of the photographs are nevertheless still realized by the individual beholders. While Mariah might think the pictures of herself and her husband under the Eiffel Tower or Big Ben are romantic, Lucy reads the fact that the photographs replicate the cultural codes of romance and of a happy family as indexical for Mariah's phoniness.

Later, Lucy pointedly asks: "Why is a picture of something real eventually more exciting than the thing itself? I did not yet know the answer to that" (121).[136] Considering Lucy's insights with regard to Mariah's photo albums, it becomes clear that what makes photographs so exciting to Lucy is the openness of the photographic signifier. It may be interpreted in various ways, which means that what it represents – a past event – is also always open to interpretation.

Annie John's Iconoclasms: Renouncing Representations of the Past

If the meanings of a photograph are eminently produced in the act of its perception, the relationship of the signifier, i.e., the photograph, and its signifieds are highly unstable, as they may change with every beholder. This suggests that an image's meanings may be altered on purpose: *Annie John*'s protagonist makes changes to a set of family photographs and thus declines the significance of the referents and their impact on her present. This scene has generally been identified as a key passage in Annie's separation from her mother.[137] At the age of fifteen, Annie has recently had a serious quarrel with her mother. Annie John senior had called her daughter a slut for talking to a group of boys in the street, which left Annie junior feeling vastly misunderstood and misjudged by her overbearing mother. As a result, Annie does not accept her mother's rebuke and displays an air

136 The consideration of the representation as more significant than its referent is later taken up again in Kincaid's *My Garden (Book):*. See chapter 3.2.

137 See, e.g., articles by Banerjee, Jackson, Murdoch, or Nasta.

of pride and indifference in rejecting her authority as a kind of punishment. She soon thereafter asks her father, a carpenter by profession, to make a trunk for her, in order to discard her mother's trunk, which she had used until then. Wanting a trunk of her own demonstrates Annie's separation from her mother, as Annie senior used her now symbolically charged trunk to leave her own parents, whom she felt oppressed by at the age of sixteen. Hence Annie's request has often been read as the beginning of her own flight from her family.[138] However, while these events might mark the onset of Annie's adolescent separation from the family's nucleus, in her request for a trunk of her own she ultimately integrates herself into the history of the family and in particular into that of her mother. Her attempt to demonstrate independence fails in that Annie here simply repeats her mother's pattern of behavior in asking for her own trunk. This entanglement of attempted separation and repetition, which unwantedly becomes a confirmation of affiliation, foreshadows Annie's further process of leaving her family – while she might want to gain independence, she cannot break the links as easily as she can carry out a performance of independence, which turns out not to be that at all.

After the altercation with her mother, it is ultimately Annie junior who suffers more from the break in the formerly close relationship, and not her mother, who now punishes Annie junior by withdrawing affection (*AJ* 103). In a combination of feeling angry, defeated, and heartbroken over the loss of the relationship, Annie falls into a depressive state that is described as an unexplainable "sickness" (*AJ* 124), which coincides with a lengthy period of rain. Although no physical ailment can be determined, Annie is too weak to leave her bed and feels pinned down by the rain that falls on the roof (*AJ* 109, 111). Since this weakness directly succeeds the quarrel between Annie and her mother, it may well be read as its psychological effect on the pubescent girl who agonizes over the changes in the family dynamics. Annie's subsequent engagement with a number of family photographs is part of this context.

Two weeks into her now delirious state, Annie is left alone by her parents, who begin to resume their regular daily activities when Annie shows no signs of improvement. As soon as she is by herself, a group of family photographs arranged in Annie's room attracts her attention, when they suddenly loom up big in front of her (*AJ* 118). The images depict important family events in Annie's life: "each illustrates one of the institutions into which Annie is in the process of being inducted as she becomes a social subject," as Marianne Hirsch explains ("Resisting Images" 258). In one photograph Annie wears her school uniform, in another she is a bridesmaid at an aunt's wedding, a third shows her father and

138 See, e.g., Jackson, Murdoch, Perry.

mother together, and a fourth depicts Annie on the day of her First Communion (*AJ* 118). While Annie is looking at them in the state of her sickness, the photographs take on a life of their own:

> The photographs, as they stood on the table, now began to blow themselves up until they touched the ceiling and then shrink back down, but to a size that I could not easily see. They did this with a special regularity, keeping beat to a music I was not privy to. Up and down they went, up and down. They did this for so long that they began to perspire quite a bit, and when they finally stopped, falling back on the table limp with exhaustion, the smell coming from them was unbearable to me. I got out of bed, gathered them up in my arms, took them over to the basin of water on the washstand, and gave them a good bath. I washed them thoroughly with soap and water, digging into all the crevices, trying, with not much success, to straighten out the creases in Aunt Mary's veil, trying, with not much success, to remove the dirt from the front of my father's trousers. When I finished, I dried them thoroughly, dusted them with talcum powder, and then laid them down in a corner covered with a blanket, so that they would be warm while they slept. (*AJ* 119–120)

After the washing, the photographs are permanently damaged, as Annie notes: "None of the people in the wedding picture, except for me, had any face left. In the picture of my mother and father, I had erased them from the waist down. In the picture of me wearing my confirmation dress, I had erased all of myself except for the shoes" (*AJ* 120). Previous analyses of this scene, such as Marianne Hirsch's or Susheila Nasta's, have concentrated on the destructive outcome of Annie's handling of the images, to then interpret this psychologically.[139] I, however, will attend to her reasons for doing so that are apparent in the specific ways in which Annie engages with the images, to then ask how Annie understands the ontology of the photographs and investigate their capacities to represent Annie's personal past.

Marianne Hirsch argues that by effectively destroying the photographs, Annie alters her perception of the past: "In washing and changing the pictures and relationships they depict, Annie can attempt to weaken the 'having been there' [Barthes] of the referent. She can attempt to intervene in her past, to rewrite her memories in favor of representations that better fit with the new life she wants to explore" (Hirsch, "Resisting Images" 260). By weakening the photographic referent, she also weakens the memory of the past (Hirsch, "Resisting Images" 263),. which creates space for new identity-constituting moments in Annie's life. As Susheila Nasta writes: "[P]erforming the deliberate action of erasure, she attempts to

139 Hirsch, for instance, proposes that Annie "acts out her fear of adult sexuality as she tries to remove the photos' bad smell, erase her parents' bodies below the waist, and rub out the stain on her father's pants and the wrinkles in the wedding veil" ("Resisting Images" 259).

posit the right to choose a future identity rather than remaining bound to an already prescribed narrative of her past" (79). I find these interpretations are convincing, particularly if we consider that when Annie first falls into her depressive state, she perceives her illness as embodied by a "black thing" in her head (*AJ* 111) that "shut[s] out all [her] memory of the things that had happened to [her]" (*AJ* 111–112). The break with her mother and Annie's contemplation of this rupture already leave her calling her former self-identification into question. Her earlier life matters little in the process of becoming an adult in which she takes on a new self.[140]

However, taking into account that Roland Barthes's conception of the photographic referent, which Hirsch employs here, is concerned with the real object that was in front of a camera at one point in the past, Annie could not possibly "weaken the 'having been there' of the referent," as Hirsch puts it, since Annie would not have access to a past reality. What she physically changes is the signifier of that referent, namely the photograph. Roland Barthes writes that "[w]hatever it grants to vision and whatever its manner, a photograph is always invisible: it is not what we see" (*Camera Lucida* 6). Annie's washing of the photographs, however, foregrounds their mediality – the vehicle of the photographic sign is made visible.

While Hirsch's and Nasta's takes on the scene in *Annie John* highlight the variety of the possible meanings that photographs may implicate, they do not attend to the processes of Annie's iconoclastic act. She does not just get up and drown the images in a washbasin with the clear objective to efface them. Rather, the photographs spring into action and engage with Annie, who initially remains a passive observer. In Annie's (albeit feverish) perception, *they* come alive; they change sizes by blowing *themselves* up, which suggests that they possess some form of agency. Then they shrink to a size so small that Annie can hardly see them anymore, which indicates their power to evade Annie's gaze. Annie clearly is not in control of the situation; on the contrary, the images decisively exclude her from their dance as Annie cannot even discern the music that seemingly provides the beat to which they change their sizes. Through their motion, the photographs moreover acquire a rather human bodily function: they begin to sweat. Just like humans would, the images are tired out by their exercise, which leaves them limp and exhausted. In this state, the photographs now give off a smell (most likely from their perspiration) that is "unbearable" to Annie and which finally incites her to take them up and give them "a good bath" in an appliance that is commonly used for human hygiene. Accordingly, it is not a will to destroy, but a feeling of disgust that prompts

140 Hirsch points out that after this episode, Annie reinvents herself by wearing a new set of clothes, speaking with a made-up accent and eventually leaving Antigua ("Resisting Images" 259).

Annie to put the images into the basin. While washing them to remove the smell, Annie suddenly perceives the photographs as three-dimensional objects when she attempts “digging into all the crevices, trying […] to straighten out the creases in Aunt Mary’s veil” (*AJ* 119–120). Here, Annie not only tries to clean the images’ contents, but significantly, her venture to straighten the creases is also an effort to eradicate the three-dimensionality. Annie attempts to make changes to the iconic sides of the photographic signs. The icon nevertheless remains stable; trousers and veil do not take on different forms, as the twice used phrase “with not much success” indicates. As little as Annie could get to the images’ referents, since they are located in the past, she cannot alter their iconic likenesses either. Her efforts remain arrested on the level of the images’ mediality. After all, she does change the photographic sign by washing and scrubbing it, but her work primarily affects the medium – the paper – and only secondarily (by way of destruction) does it affect the sign the medium transports.

Although Annie is unable to take hold of the photographic icons, they nevertheless appear before her. In their movement, the images transform so that the icons begin to dominate. Susan Sontag clarifies the difference between the photographic representation and its referent in terms of power relations: “a photograph is not like its subject, a homage to the subject. It is part of, an extension of the subject; and a potent means of acquiring it, of gaining control over it” (155). In Annie’s case, these roles are reversed when the photographs assail Annie and exert power over the girl. When she then washes the photographs, she pushes back against the power of the photographs that physically emerge into the room. Their human-like perspiration, which implies a corporeality, offends her as much as their three-dimensionality. Hence, in her washing of the images the intended goal is not to destroy them (although that is the inevitable outcome), but to return them to their former immobile and two-dimensional state of the photographic medium.

In immersing the photographs in the washbasin, Annie moreover repeats the act of producing photographic prints. During the process of printing, the developer bath brings forth the image on the previously (seemingly) blank paper. When Annie washes her family’s photographs, she, too, creates images. Although these are not new images, Annie’s imitation of the developer bath changes the photographs so that afterwards they show different versions of their former selves. Annie does not have blank canvases to work with to produce radically new versions of the photographs and, with that, of herself, as Hirsch puts it, but she is still able to create different versions, which reduces the visibility of the other family members and lets her own likeness appear at the center of some of the photographs.

After washing the photographs, Annie dries them, dusts them with talcum powder – a hygienic practice to prevent perspiration – and cradles them in a blanket to keep them warm. Again, these are not destructive acts, but rather those of mothering. The use of an antiperspirant suggests the intention to prevent such an episode from happening again and thus rather takes care of the photographs than purposefully damages them. It also indicates that Annie intends a future use for the photographs and did not just instrumentalize them to act out a singular performance of separation and independence.

Ultimately, by handling the photographs the way she does, Annie changes and partly obliterates them. She does not, however, as Hirsch and Nasta have argued, directly act upon the photographic referents to alter her past. Rather, as shown above, in acting on the photographs Annie emphasizes their mediality. When the images' icons seemingly come to life, Annie pushes back against their presence and returns them to their medium – the two-dimensional paper. Furthermore, even though Annie tries to make changes to the referents' likenesses, she is unable to do so. She does, however, effectively change their medium and consequently their signifiers so that they not do not indexically point to their referents by way of the production process of photographs (see Peirce 159, Dubois 40 – 50), but now the altered surfaces of the images function as a trace that again points to the production process of the images, though this time to that of their alteration – the "second developer bath" that Annie gave them. Thus, the photographs' significations are redirected from family events and Annie's daughterly role in this social structure to Annie's washing of them and to her self-determination.[141] The photographic sign now still works in the same way – indexically – but Annie's intervention on the plane of the medium has changed the signifieds. Perceiving photographs as medium that function as signifiers, questioning the images' referential power, and thereby destabilizing a conventionally direct connection of signifier

141 By erasing all faces except for her own in one picture and by expunging her confirmation outfit except for her shoes, which she had chosen herself and which had become a bone of contention, as her mother thought them inappropriate (*AJ* 119), Annie emphasizes her autonomy from her family. However, to be clear, while I argue that Annie takes a step towards self-determination and independence from her family in her treatment of the photographs, I do not maintain that she achieves it in this single performance. Several points would contradict this stance. First, she does not act as a self-determined subject in that she is in a delirious state, when she engages the photographs. Second, she does not choose to engage with them, but when they approach her, she initially remains passive until provoked by the smell the images give off. Third, the separation from her family remains incomplete throughout the text. While steps towards Annie's independence characterize her narrative, and although the book ends with her physically leaving Antigua to attend nursing school in England, Annie feels both relieved and devastated; just as her mother makes clear that their complicated relationship cannot be concluded by geographical distance: "It doesn't matter what you do or where you go, I'll always be your mother and this will always be your home" (*AJ* 147).

and referent provides space for new significations and finally allows Annie to produce a representation of her past – one in which she is not a dutiful daughter as the family photographs might suggest, but which establishes her as independent, and on the basis of which she now may proceed to negotiate her identity in the present and in the future on her own terms. I read Annie John's changing of family photographs as exemplary for the way in which Kincaid's works engage with historical representations. Through repetition (here of the developer bath), something is uncovered that was already inherent but obscured in and by the representation. Changes to the medium, such as Annie John's altering of the photographs, change the foci of perception and reveal versions of history that are not radically *new*, but radically *different* in that they now center on previously hidden participants in this history.

Such visual practices are employed by the characters of Kincaid's texts not only concerning their personal histories, but also with regard to collective colonial history. Visual representations of colonial rule are discarded, destroyed, and altered, which in each case entails meaning-making processes that do not conform to colonial ideals.

Annie John, for instance, rejoices about her new notebooks, which have a cover of black and white blotches mixed up together: "so glad I was to get rid of my old notebooks, which had on their covers a picture of a wrinkled-up woman wearing a crown on her head and a neckful and armsful of diamonds and pearls" (*AJ* 40). Shalene Vasquez identifies the portrait as that of Queen Victoria, "one of the most influential figureheads of colonial Antigua" (Vasquez 34). As such a figurehead, the image of the queen on a school notebook signifies the colonial indoctrination that was part of the educational system in British colonies. Annie's happy discarding of the books and her preference for the more neutral book covers then emphasizes her disregard of the colonial rule.[142] Significantly, Annie does not even mention the person portrayed by name, which once more highlights her indifference to the supposed figure of authority. Similar to her erasure of the family photographs, in throwing out the notebooks with the queen on the covers Annie here refuses the visual representations of history to make space for a less overbearing present on more neutral terms.

More unequivocally even, the protagonist of *The Autobiography of My Mother*, Xuela, breaks a decorative porcelain plate when she is a young girl. The plate displays an image of a landscape: "This picture was nothing but a field full of grass

142 One could even argue that a mix of black and white spots represents the exact opposite of colonial rule, as here no color prevails over the other but they are intermingled – be it literally with regard to blotches of color on a book cover, or metaphorically with regard to human skin colors.

and flowers on a sunny day, but it had an atmosphere of secret abundance, happiness and tranquility; underneath it was written in gold letters the one word HEAVEN. Of course it was not a picture of heaven at all; it was a picture of the English countryside idealized" (*ABM* 9). The owner of the plate, Eunice, is unaware of the idealization of England, according to Xuela, she indeed thinks that the image depicts a heaven that promises a carefree and worry-free afterlife (*ABM* 9). For Eunice, the plate succeeded in conveying English ideals and imagination in the colonized space, when it conjoins the English countryside by way of iconicity with the Christian promise of blissful eternity through the inscription, which connects the symbolic paradisical meaning to the English countryside.

Xuela does not intentionally drop the plate to break it, but neither does she apologize for it. Eunice is devastated by the loss of the token and physically punishes the girl who is in her care, but still Xuela refuses to ask for forgiveness. Since at the time Xuela herself was unaware of the picture's colonizing agenda, her dropping of the plate is hardly an intentional performance of resistance against colonial rule. Nevertheless, this iconoclastic act occurs on the first few pages of the book and introduces the protagonist's deviant behavior and her instinctive rebelliousness against English rule, which are Xuela's central character traits that feature throughout the narrative of her life. Breaking the plate, Xuela destroys both the symbolic signifier of a Christian heaven as well as an idealized, idolized, and iconic signifier of England. She breaks not only the physical object, but in doing so, she also severs the connection of heaven and England which are linked in the picture. Ultimately, the breaking of the plate performs a repudiation of the colonizer's power over the imagery of the metaphysical concept of heaven as well as the paradisical attributions to the glorified "mother country."

Such colonialist iconography is moreover rejected not by destruction but by alteration in *Annie John,* when the protagonist reassigns meaning to an image of Christopher Columbus in her school book. During class, Annie skips ahead in her history book and is immediately fascinated by one of the rare color pictures in the book: it shows Columbus sitting at the bottom of a ship with his hands and feet shackled, looking rather miserable after having been arrested by Francisco de Bobadilla for his poor government of Hispaniola. "How I loved this picture – to see the usually triumphant Columbus, brought so low," Annie muses (78). Seeing the image of Columbus in this position prompts Annie to think of her mother's rejoicing at her grandfather's physical demise. Annie John senior has had a quarrelsome relationship with her own father, Pa Chess, and spitefully relishes his inability to move about as he pleases. Pa Chess is shackled by a stiffness of his limbs, which he has acquired with his old age. Venomously, Annie John senior remarks: "So the great man can no longer just get up and go. How I would love to see his face now!" (78). Annie recognizes the analogy between her autocratic grandfather Pa

Chess and the colonizer Christopher Columbus, their shared physical inhibitions, and the delight of those previously oppressed by the two men. Fittingly, Annie then draws over the image's title "Columbus in Chains" and replaces the descriptive caption with her mother's dictum "The Great Man Can No Longer Get Up and Go" (*AJ* 78). By changing the caption, Annie effectively changes the meanings of the image. It now does not show a "great" European colonizer who simply was imprisoned at one point in his life, i. e., an authoritative historical figure; but by aligning him with her own disliked grandfather and appropriating him into her own genealogy, Annie removes Columbus's historicity. Christopher Columbus now is just an powerless man to be laughed at.

Alisa Braithwaite notes that "[b]y placing writing on a page that is already inscribed, [Annie John] changes the location of writing itself so that she can start her narrative wherever she desires. [...] There is no originary blankness on which the writer writes. The page is a text, a location already occupied upon which the writer negotiates her narrative" (151). Indeed, Annie does not write on blank space and her alteration of the image's inscription does not – strictly speaking – change the image itself. As with the family photographs, Annie's intervention results in a redirection of significations, which allows for different meanings to emerge. Like her personal past, embodied by the family photographs, the significance of colonial history can also be changed through visual practices, although it nevertheless keeps on determining Annie's present. Yet its weight can be changed, which is what Annie does in her modification of the image's caption. By altering the inscription, which adds, to the image's iconic meaning (an illustration of a moment in a man's life), a symbolic one (every ruler can be brought ever so low just like a common man, even the most famous one), Annie gains the agency to move from being a colonial subject to a more self-determined subject.[143] Again, and like Xuela, she refuses colonial iconography and its prescriptive meanings and thus affirms her own agency to be able to attribute meanings herself.

Jamaica Kincaid's characters deny fixation as a characteristic of images. Neither photographs nor other visual representations may fix the past – be it personal or collective. The characters claim the agency to interpret the images they encounter themselves and reject being governed by them and their representations of the past. It is this variability in the relationship of signifier and signified that opens the images to the possibility of altering perceptions of the past. In Kincaid's

143 Adlai Murdoch here even reads a validation of Annie's identity: "By taking on established authority from one culture and imposing her own (written) will upon it, Annie in effect prefigures her eventual treatment of her mother's authority – by force of will, she rewrites the maternal paradigm, the clash of cultures signifying a portent of the means by which she engages in validating her own identity" ("Severing" 337).

later novels, *The Autobiography of My Mother* and *Mr. Potter*, the instability of referent and representation becomes even more productive when actual photographs on the book covers create notions of authenticity for the novels' fictional accounts of the past. Here, it is not the characters but the novels' readers who discover the unfixability of the past in images, which then allows for new knowledge of the past to emerge, as I will show in chapter 2.3 "Photographic Product(ion)s of History." Likewise, if the image is not a reliable medium to access the past, this also opens them to the possibility of productively engaging the past, as Lucy does in her refusal to be framed by others and in turn by taking photographs herself.

Lucy's Photographs: Appropriations and Distancing in the Making of the Present

Whereas Annie John negotiates the meanings of her personal past as well as her current subject position by altering family photographs, Lucy's engagement with paintings and photographs highlights her liminal position as a newly arrived immigrant in the United States. In the process of negotiating new spaces and her positions in them, Lucy takes up a camera. I read Lucy's engagements with visuality as signifying on her subject positions, which are worked out in the negotiations of colonial visual topoi.

The novel *Lucy* sets out with the arrival of the nineteen-year-old in a chapter entitled "Poor Visitor." In the course of the novel, the text details Lucy's first year in the United States – both her joys and pains in navigating the strange new world around her and rejoicing at having left behind her former home in the Caribbean, while simultaneously being plagued by feelings of homesickness. In her former life in the Caribbean, Lucy objected to being forced into social roles, especially that of the dutiful daughter. With the move, she hopes to escape "the harsh judgments […] by people whose only power to do so was that they had known [her] from the moment [she] was born" (*Lucy* 51). However, once in the United States, Lucy has to realize that even where nobody knows her, she is still not free from being framed by others. While in the United States she is not her mother's daughter, she now is an immigrant girl, a nursing student, the help, a girl from "the islands," or simply "the girl" (*Lucy* 58). Leaving the Caribbean lets her escape neither her personal past, nor a shared colonial history, which prevails in the form of casual racism (see, for instance, *Lucy* 56–58). Lucy traveled to the United States to work as a nursemaid for a wealthy American couple and to pursue a degree in nursing in night school.[144] Uninterested in either of these occupations, Lucy hoped in coming to the United States to leave behind her former home in the Caribbean and with that her former life, i.e., her personal past. Putting a geographic distance between herself and her mother, who dominates Lucy's thoughts of home, however, is not enough for the girl to (re)invent herself on her own terms, as the longing for home and (more or less) loved ones only emerges in their absence. Lucy cannot achieve individuation and independence, because her past follows her into her new life in her memories. Simultaneously, she never fully arrives in her newly adopted country, as she is unwilling to fully assimilate, not least because she is entangled in the continual process of disengaging from her former home.

Lucy's physical move from the Caribbean to the United States presents a caesura that splits her life into two parts: The realization that she has left the tropics "entered [Lucy's] life like a flow of water dividing formerly dry and solid ground, creating two banks, one of which was [her] past [...] the other [her] future, a gray blank, an overcast seascape on which rain was falling and no boats were in sight" (*Lucy* 5–6). Lucy separates her life before and after her crossing of the ocean, and although she is eager to leave the Caribbean, what lies before her is not a positive outlook, but rather a bleak, unsteady future. While she has crossed the ocean physically, metaphorically she is still at sea, caught in a state of in-betweenness: torn between two banks, she sees "the present take the shape of [her] past" (*Lucy* 90). In her state of transition, Lucy carries the sea – the space of crossing – within herself, which becomes a constant in her life. Metaphorically illustrating the poetics of impermanence in Kincaid's works, the stability of two fixed poles is thus dissolved in *Lucy*. Instead of choosing between the binaries home/away and past/present, Lucy inhabits the interstice between the extremes and denies their assumed polarities. A seesaw motion is also played out in the text's narrative strategy of alternating between Lucy's present in the United States and her past, by associatively slipping into memories of her former life in the Caribbean. The narration thus churns between temporal levels and narrative spaces, emulating a movement of being at sea. Especially Lucy's photographic practices highlight this position. She appropriates her environs by photographing them, claiming moments and situations as her own, just as she conversely distances herself from

---

144 Indeed, *Lucy* is often read as a sequel to *Annie John*, as the protagonists' biographies are highly similar. *Lucy* appears to pick up where *Annie John* leaves off, with the protagonists moving away from their former homes in the Caribbean and towards the pursuit of a nursing degree – although Annie leaves for the UK, while Lucy arrives in the United States. See, e.g., Bouson 67; Simmons, *Jamaica Kincaid*, 120; Jackson, or Bolaki. I, however, regard Annie and Lucy as similar but separate characters in separate novels. The discussion of continuity consequently does not affect my reading of photographic practices in *Annie John* and *Lucy*.

her surroundings by putting a camera between them and herself. I argue that this back-and-forth motion is at the heart of Lucy's positioning and self-identification after her move to the United States.

Lucy's desire for individuality and independence is further evident, for instance, in her rejection of Mariah's well-intended suggestions to read feminist works such as Simone de Beauvoir's *The Second Sex* (*Lucy* 131–132). Since she regards her troubles her very own and is unwilling to be classed with *any* social group, Lucy refuses to recognize them as those of all women (see *Lucy* 139). She chooses not to identify with feminist agendas because that, again, would confine her to a role. Hence Lucy cannot get past the first sentence of Beauvoir's text: "Woman? Very simply, say the fanciers of the simple formulas: she is a womb, an ovary; she is a female – this word is sufficient to define her" (*Lucy* 132). Beauvoir's ironic reference to a biological definition of women leaves aside any individual traits and lumps together a very large group of people based on their having certain inner organs, which starkly contradicts Lucy's desire for autonomy. Furthermore, Nancy Chick observes that Mariah's advice could not possibly serve Lucy because it "resembles the argument espoused by many women of color that the Feminist Movement of the 1960s and 1970s has centered on the concerns of white, middle-class women, ignoring the perspectives of minority women" (93–94) – although, I add, Lucy would in all likeliness also refuse the category of "minority women."

In Lucy's life in the United States, one instance of being confined to a social position not only frames Lucy metaphorically, but quite literally frames her in an image: her boyfriend Paul makes her a present of "a photograph he had taken of me standing over a boiling pot of food. In the picture I was naked from the waist up; a piece of cloth, wrapped around me, covered me from the waist down" (*Lucy* 155). The image recalls exoticizing depictions of "island girls," presenting them both as innocent (carrying out everyday tasks, such as cooking and being unfazed by their own nakedness) and as sexualized (showing them bare-breasted). In this, such an image not only frames the depicted in the desires of the onlooker, but also deeply impinges on their privacy. A famous namesake of Lucy's boyfriend, who is also well known for such images of "island girls," is the French postimpressionist painter Paul Gauguin, who visited in Tahiti from 1891 to 1893 and after a brief stay in France again traveled to Tahiti to then move on to the Marquesas Islands in 1895, where he lived until his death in 1903.[145]

145 Ironically, the editors of *Lucy* chose Gauguin's painting *Poèmes barbares* (1896) to illustrate the novel's cover. The choice of an image of a brown-skinned girl, only clad in a piece of cloth from the waist down, is an ironic one: the placement suggests an association of the image with Lucy, as pro-

Lucy rejects the photograph taken by her boyfriend Paul and its repetition of Gauguin's framing and exoticization in depicting the naked racial Other. She discerns: "That was the moment he [Paul] got the idea he possessed me in a certain way, and that was the moment I grew tired of him" (*Lucy* 155). Significantly, Lucy here connects imagery and possession. The painter Gauguin had himself observed about his work on the women of Tahiti: "I was aware that my exercise as a painter was like a close study of the subject's interior life, like taking physical possession, like a tacit and urging plea, like an absolute and definite conquest" (*Noa Noa* 47–48).[146] Paul's image of Lucy, though in a different visual medium, hardly differs from Gauguin's painting in terms of appropriative practices, since, as Susan Sontag notes: "[t]o photograph is to appropriate the thing photographed. It means putting oneself into a certain relation to the world that feels like knowledge – and, therefore, like power" (4). In this sense, both photography and painting constitute acts of colonization as the persons depicted are taken into possession by the onlooker and artist through the visual renditions of their likenesses.

Paul's possessive act is moreover twofold, as Lucy grows tired of him not when he snaps the picture, but when he presents it to her and thus demonstrates his ownership of her likeness and by extension of her body, as he his free to display it to the external world although the image was taken in an intimate situation.

Lucy reacted to being confined to the social role of the "daughter" in the Caribbean by physically removing herself and to attempting to do so emotionally, as well. Similarly, soon after receiving Paul's gift of the photograph, Lucy leaves Paul and ends the relationship. Lucy not only rejects being photographed and in this way being possessed by Paul, but like Annie John, she also objects to being presented in a manner that is not of her own choosing, i. e., to being put on display as the exotic Other by Paul's photograph. This once more elucidates Lucy's refusal to be fixed in representations by others – both in life in general and more specifically through visual depiction. Moreover, if one reads Paul's photograph as a repetition of Gauguin's colonial paintings of "exotic girls," as the artists' congruence of names further suggests, then Lucy's rejection of the image not only pertains to her own framing in Paul's picture; the text thus engages with the topos of the half-naked girl in colonial visual arts, which is both exposed and rebutted by Lucy's response.

---

tagonist and eponym of the novel (see chapter 3.3 for a discussion of image positions and assumed ties to characters within the texts), yet Lucy herself would vehemently reject such a portrayal.

146 French original: "J'eus conscience que mon examen de peintre comportait, avec une profonde étude de la vie intérieure du modèle, comme une prise de possession physique, comme une sollicitation tacite et presante, comme une conquête absolue et definitive" (Gauguin, *Noa Noa* 47–48). Translation mine since the one by O. F. Theis (1919) is inaccurate and incomplete (see also Gauguin, *Noa Noa* (1919) 33).

In the course of the novel, Lucy reverses the dynamics of appropriation by turning her own gaze onto others. Her refusal to be the object of expectations and desires is followed by her own repetition of such colonizing gestures. Kincaid's text thus lays bare the colonial legacies that prevail in everyday interactions, as Lucy exposes them in performing them. She positions Americans as racial and cultural Others to herself, who can be framed in the same exoticizing ways as white colonizers did (and do) when travelling to non-Western countries.[147] This move is signaled by Lucy's identification with the painter Paul Gauguin, especially after the rejection of her own image in Gauguin-style.

Lucy reads Paul Gauguin's life as parallel to her own when Mariah takes her to a Gauguin exhibition. Moira Ferguson muses that Mariah might have taken her young friend to the exhibition with the intention to present her to paintings of people like her ("Lucy" 247). Lucy's interest, however, is not directed at the paintings, but towards the painter Gauguin himself,

> a French man, who had gone halfway across the world to live and had painted pictures of the people he found living there. He had been a banker living a comfortable life with his wife and children, but that did not make him happy; eventually he left them and went to the opposite part of the world, where he was happier. [...] [I]mmediately I identified with the yearnings of this man; I understood finding the place you are born in an unbearable prison. (*Lucy* 95)

Lucy here perceives and identifies with the painter's discontentment and his attempt to geographically move away from it.[148] Indeed, before leaving for Tahiti, Gauguin felt himself and his art misunderstood by the European art scene of the late nineteenth century, as one of his letters to Vincent van Gogh shows: "Alas, I see myself condemned to be less and less understood, and I must hold fast to following my way *alone*, to drag out an existence without a family like a pariah. So the solitude in the woods seems to me in the future to be a new and almost dreamed-of paradise. The savage will return to savagery!" (Gauguin, "My Dear Vincent," emphasis in original).[149] Gauguin's travels in this sense present a

---

147 See Kincaid's *A Small Place* for a comparison of colonialists up to the twentieth century and current neocolonialists who travel to Caribbean countries as tourists.

148 Bénédicte Ledent even reads Lucy's voyage as "an echo of Columbus's original confrontation with the uncharted realities of a new world" (53).

149 However, it stands to reason that Gauguin left France not to live a life more in accord with nature, but as cheaper than in France, as he was constantly strapped for money. Moreover, while he may have sought to perfect his art by immersing himself in an unfamiliar space and culture, he did count on a European demand for paintings of the exotic unknown (see Thomson 131).

flight from his current identity as an unsuccessful painter towards a self-invention as "savage."[150]

Lucy's own desires to disengage from her old life structurally mirror Gauguin's reasons to travel. Moreover, Lucy will eventually also repeat Gauguin's picturing of the natives of the adopted country when she later takes up photography. Lucy thus chooses to identify with the subject position of the onlooker, the active party, the one who frames others, and hence with the one who performs colonizing gestures, rather than with the "island girl" who is the object of gazes and images. In her own performances of visual power, Lucy thus highlights and reverses prevailing racial relations of colonizer and colonized.

Jennifer Nichols asserts that "Lucy's visit to [the] Gauguin exhibit helps her begin to see [...] herself as much of a creator as a character – rather than a completed project that may be represented but never altered" (195).[151] While Lucy's identification with Gauguin foreshadows her later role as a creator, it is not immediately incited by the visit to the painter's exhibition. Another museum visit and the catalogue of a photography exhibition come before Lucy's decision to create herself. Here, she is most intrigued by "photographs of ordinary people in a countryside doing ordinary things, but for a reason that was not at all clear to [her] the people and the things they were doing looked extraordinary – as if these people and these things had not existed before" (*Lucy* 115). Assuming that these photographs show people and things in an *American* countryside, the images could be the counterparts to Gauguin's renditions of Polynesia, when in both the ordinary occupations of the Other are unfamiliar and hence exotic to the foreign onlooker and thus become worthy of attention. More importantly, Lucy perceives the people

---

150 *Noa Noa,* Gauguin's written account of his stay in Tahiti, must in any case be taken with a grain of salt. As Belinda Thomson or Eckhard Hollmann have pointed out, for instance, Gauguin did not intend to provide a travel report for its own sake. Rather, his intention was provide the reader with access to his artistry as much as it was to make money from its publication (Hollmann 42). Gauguin's extensive "borrowing" from both Pierre Loti's *Le mariage de Loti* (1880), a widely sold travelogue of a French soldiers' sexual relations with a Tahitian girl, and Jacques-Antoine Moerenhout's *Voyages aux îles du Grand Océan* (1837) (see Hollmann 27, 48, and Thomson 156) moreover suggest that Gauguin wrote to please an audience, drawing on already established narratives of Polynesia.

151 Nichols derives this interpretation from Kincaid commenting on Gauguin that her interest in him "refracted onto the canvas of Kincaid's novel through Lucy's reaction to his work and his life. His brief role in the novel [...] sheds light on Kincaid's textual practices of outing 'purity' as a deeply-rooted seed in American culture and using her protagonist to interrogate its conceits" (Nichols 195). While Nichols focuses on extratextual sources to interpret the occurrence of Gauguin, looking at the primary text *Lucy,* I nevertheless come to a similar conclusion, considering Lucy's desire for change and self-making.

and things as only brought into being by their visual representations. Their existence is not only attested to by the photographs; in their perceivable and current forms they are rather *made* by the photographs when they exhibit their ordinariness as extraordinary. In this sense, photography indeed is creation and not simply the depiction of something that already existed. Reading this in parallel to the medial processing of the past into history, be it via text or visual media, Lucy's realization lays bare the importance of media in the creation of historical consciousness.

Lucy's conception of photography here picks up on twentieth-century discourses on photography as art in which photographs are not merely thought as mimetic depictions of a reality (see Dubois 31–40). Yet some in this tradition, such as Pierre Bourdieu, regard a photograph as the result of a random choice in selection of the subject and situation and accordingly as an edited rendition of reality (73). In this understanding of photography, the dark room is then not a space of neutral reproduction, but much rather one in which effects on photographs are consciously and purposefully created (Dubois 36). Lucy's interest in photography thus pertains to its creative properties as much as to its mimetic ones. Her encounter with photographs that render extraordinary both ordinary people and things finally incites her to buy herself a camera (*Lucy* 115–116).

The photographs Lucy initially takes are of her immediate environment: of the family she lives with and the space she occupies in their apartment. Attempting to imitate the mood of the images in her book of photography, Lucy takes pictures of Mariah while cooking and of the children eating marshmallows. Just as in the photographs she tries to emulate, and as in Gauguin's paintings, the people in Lucy's images are occupied with everyday tasks. Moreover, Mariah and her children had initially seemed very strange to Lucy (see, for instance, 12–13), so that the very common situations of preparing a meal or consuming sweets may appear somewhat unfamiliar to Lucy in her perception of the American family. The meal that Mariah cooks in the photograph is "an elaborate preparation of chicken and vegetables slowly cooked in red wine" (*Lucy* 120–121), which may well be the French dish coq au vin and which would certainly seem foreign to Lucy, as her incomprehension of "French-cut green beans" (*Lucy* 14) suggests. In picturing Mariah cooking, Lucy repeats Paul's image of herself, which she had so fiercely rejected, as well as Paul Gauguin's paintings of his racial and cultural Others. In reading the taking of photographs as colonizing acts, Lucy's picturing of Mariah moreover reverses and revises stereotypical racial power relations when the brown girl turns her gaze and her camera at white middle-class Americans.

It is noteworthy that Lucy takes photographs of others, but not of herself, as perhaps could be expected, considering her desire for self-making. bell hooks notes that "[t]he camera was the central instrument by which blacks could disprove representations of us created by white folks. The degrading images of blackness that emerged from racist white imaginations [...] could be countered by 'true-to-life' images" ("Photography" 48). However, when Lucy takes up the camera, she is not invested in correcting representations of herself as they might be perceived by others or in producing a counter image to Paul's exoticizing photograph of her. Lucy does not project her own image into the world, but rather takes hold of a world that is still alien to her and to makes it her own by photographing it.

Other early objects of photographic interest besides her employers' family are her own possessions in her room, such as her "dresser top with [her] dirty panties and lipstick, an unused sanitary napkin, and an open pocketbook scattered about; [...] a necklace made of strange seeds, which [she] had bought from a woman in the street; [...] of a vase [she] had bought" (*Lucy* 120–121). Lucy here documents her cluttered belongings as one likely would only leave them in one's private space, with dirty underwear and hygienic products on display. Hence, Lucy's scattered array of personal items may be read as Lucy claiming her own private space and thereby a degree of independence. Read as indexical signs of Lucy's progress in disengaging from external expectations, her way of displaying the very ordinary items would then turn them into extraordinary objects. They are Lucy's very own, chosen by her, as the necklace and the vase, and put on display by her. Her photographing them yet again signals their extraordinariness. Lucy in this way puts this form of self-expression on display when she also places the photographs of her room within her room for herself to look at (*Lucy* 120–121). Rather than creating representations of herself for an outside world, Lucy here produces representations of her own space and choices for herself. "Cameras establish an inferential relation to the present (reality is known by its traces), provide an instantly retroactive view of experience. Photographs give mock forms of possession: of the past, the present, even the future," Susan Sontag writes (167). Accordingly, through photography, Lucy not only claims space and choices; more than that, she can claim her own past, present, and future.

Eager to detach herself from anyone's claims on her body and on her life, Lucy realizes how much her move to the United States is aiding her in her endeavor. After the first few months in this new country, she notes: "I could now look back at the winter. It was my past, so to speak, my first real past – a past that was my own and over which I had the final word" (23). It is this final word that is also at the center of Lucy's photographic documentation of her room. The images present Lucy's final word fourfold: first, in the acquisition of the objects; second, in how they are placed in the room; third, in that and how they are photographed; and fourth, in the display of the photographs that present her previous decisions. Susheila Nasta also reads Lucy's visual practices as liberation from her former self through the creation of a photographic chronicle of her own life (79–80).[152] And although, as I have shown above, Lucy is less concerned with projecting representations of herself towards an outside world, bell hooks's assessment that "[m]ore than any other image-making tool, it [the camera] offered African Americans [and here an Afro-Caribbean immigrant] disempowered by white culture a way to empower ourselves through representation" ("Photography" 48) still holds in Lucy's case. The images of her possessions represent parts of Lucy's new past on her own terms, but much more than that they represent her process of actively taking control of her own life: when photography allows Lucy to claim her present spaces and choices as her own, then "what photography supplies is not only a record of the past but a new way of dealing with the present" (Sontag 166).

Later, Lucy's photographic interest leaves the domestic space of Mariah and Lewis's home, just as Lucy herself leaves the family in search of yet more independence. Increasingly seeking life outside the home, Lucy had begun to frequent parks and wander the city's streets. Eventually, she detaches herself from Mariah, too, whom she had come to love as a friend and surrogate mother figure, and Lucy takes up work as a secretary and general help for a studio photographer. Together with her friend Peggy, she rents an apartment and thus departs from Mariah and her family, who had for the past year provided a home to Lucy. She decides not to leave a forwarding address in order to be unreachable and thus repeats the departure from home that she had performed at the outset of the novel when she left her mother in the Caribbean.

Leaving the domestic space turns Lucy into a walker of the city. In the way of a modern-day flâneuse, she discovers the city, looking at men in parks, trying to read their sexual prowess on their bodies by studying their physiognomies without ever directly approaching anyone (*Lucy* 88–89).[153] Walter Benjamin describes this notion that faces might reveal the lives of those to whom they belong as the "phan-

---

152 Moira Ferguson, in her reading of Lucy's photographing, emphasizes the girl's detachment from her mother: "The camera proffers the illusion that material reality can be controlled when Lucy blocks and reverses her mother's gaze across the Caribbean and the Pacific. She uses the camera through which she gazes unrestrictedly herself to redeploy her mother's watchfulness. Scopic discoveries substitute for the unpleasant visual images rendered by her memories of the past. With a camera, memories can be rearranged, boundaries elided; photos can be used to manipulate and conceal the real" ("Lucy" 248–249). While this is a compelling reading, and while Lucy certainly feels her mother's overbearing power even when geographically distanced from her, I still argue that Lucy here attempts to escape not only her mother's gaze, but society's as a whole, both in the Caribbean and in the United States, as I have shown with regard to Paul's exoticizing gaze and Mariah's imposing feminist agenda.

153 For a general discussion of bodies as media and their being "read," see chapter 4, "Embodied Enactments of History."

tasmagoria of the flâneur" (*Arcades* 429). Although the categories – "profession," "ancestry," and "character" – that that Benjamin's flaneur might seek out in unfamiliar faces become sexual abilities in the gaze of the young woman Lucy, structurally she nevertheless qualifies as a somewhat flaneuring figure, especially when it comes to Benjamin's dialectic of the flânerie of both being at the center of the gaze and disappearing in the crowd (see *Arcades* 420), which may well be read in Lucy's demeanor when she walks the city. Flânerie is a praxis that is highly comparable with Lucy's in-between position centered on both looking and being looked at, both being in constant motion and being without a destination. This dialectic goes hand in hand with "[t]he peculiar irresolution of the flâneur," which Benjamin describes as a condition of doubtfulness: "Just as waiting seems to be the proper state of the impassive thinker, doubt appears to be that of the flâneur" (*Arcades* 425). Attempting to shed her former self, Lucy is also unsure about who to be or who to become, in doubt about whether she really can leave her past behind. While a development is suggested by the progression of chapter titles, from the first entitled "Poor Visitor" to the last, which carries the protagonist's first name, "Lucy," the girl at the center of the novel does not attain a new or profound sense of selfhood or individuation. At the beginning of the final chapter, she muses: "I had been a girl of whom certain things were expected, none of them too bad: a career as a nurse [...]; a sense of duty to my parents; obedience to the law and worship of convention. But in one year of being away from home, that girl had gone out of existence. The person I had become I did not know very well" (133). More than ever, Lucy finds herself in a state of limbo, just as she had been a year before when she entered the United States. Lucy is still "at sea" and "caught between two banks," though one of them begins to fade into the background. While she was able to free herself from claims on her and obligations she detested, this did not work to create a whole new self; rather, it affirmed her undecidedness.

As I argue in the following, last part of this section on *Lucy*, her being in constant doubt about how and who to be is also expressed by Lucy's photographic endeavors outside the domestic realm, when her subjects change from things to human beings. She muses: "I most liked to take pictures of individuals, just scenes of people walking about, hurrying to somewhere. I did not know them, and I did not care to" (*Lucy* 160). With her camera, Lucy here observes the "crowd" that is also constitutive of flânerie.[154] Susan Sontag explains that "photography first [came] into its own as an extension of the eyes of the middle-class *flâneur* [...].

---

154 Benjamin describes the "masses" as "stretch[ing] before the flâneur as a veil: [...] they efface all traces of the individual: they are the newest asylum for the reprobate and the prescript. – Finally, within the labyrinth of the city, the masses are the newest and most inscrutable labyrinth" (*Arcades* 446).

The photographer is an armed version of the solitary walker reconnoitering, stalking, cruising the urban inferno, the voyeuristic stroller who discovers the city as a landscape of voluptuous extremes" (55). Lucy's photographic interests thus turn to people who are – like her – in a state of transition, which she then freezes into unmoving permanence in her photographs.

In *The Tourist Gaze*, John Urry determines that "[t]he strolling flâneur was a forerunner of the twentieth-century tourist and in particular of activity which has in a way become emblematic of the tourist: the democratized taking of photographs – of being seen and recorded, and of seeing others and recording them" (127). Although not a tourist herself,[155] with her camera and her way of perceiving her surroundings, Lucy certainly employs a tourist gaze in her heightened visual awareness.[156] This already becomes apparent at Lucy's arrival in the United States. On the way from the airport to the city, "someone would single out to [Lucy] a famous building, an important street, a park, a bridge that when built was thought to be a spectacle" (*Lucy* 3). The sights are already known to Lucy. Before arriving, she had daydreamed that seeing them and experiencing them herself would provide her with a happiness that would counter the unhappiness she felt at home (see *Lucy* 3). Lucy here displays a tourist's anticipation, which is "constructed and sustained through a variety of non-tourist practices, such as film, TV, literature" (Urry 3). However, according to Urry, since "satisfaction stems from anticipation, from imaginative pleasure-seeking, [...] 'reality' rarely provides the perfect pleasures encountered in daydreams" (13). Lucy's expectations are, moreover, disillusioned during her arrival and when she first sets sight on tourist sites. "Now that I saw these places, they looked ordinary, dirty, [...] and it occurred to me that I could not be the only person in the world for whom they were a fixture of fantasy," Lucy realizes (4).[157] Lucy's tourist gaze emphasizes her status as a vis-

155 Urry conceives of the tourist as a traveler who intends to return to a fixed space of "home" after a predetermined period of time (see Urry 1–3), while Lucy had come to the United States to remain there indefinitely.

156 The tourist gaze is a way of looking, which may have originated with the classic tourist but is not necessarily tied to the figure anymore and has come to be attached to a way of approaching one's environs in any situation: "What I have termed the 'tourist gaze' is increasingly bound up with, and is partly indistinguishable from, all sorts of other social and cultural practices. This has the effect, as 'tourism' per se declines in specificity, of universalizing the tourist gaze – people are much of the time 'tourists' whether they like it or not" (Urry 72).

157 Lucy's disappointment regarding the realities of the country she has visited once again presents a similarity to Paul Gauguin's travel trajectories and his perceptions of Tahiti: both Lucy and Gauguin were looking from the land of their dreams – or rather of the medialized representations of their respective times – and are unable to find them in their realities. Paul Gauguin had meant to travel to a Tahiti that, to his disappointment, did not exist anymore. In his travel account *Noa*

itor at the outset of the novel as much as it highlights her disillusionment at finding a perfect new life in her new country.

While not a tourist per se, Lucy keeps touring the city with her camera, the tourist's essential tool, which lets her visit new spaces without entering into more intimate relationships with them or the people who inhabit them. As Lucy notes, she does not even care to know them (160). Susan Sontag describes how relating to unfamiliar surroundings can be eased by taking pictures, as it may have a calming effect on the photographer: "The very activity of taking pictures is soothing, and assuages general feelings of disorientation that are likely to be exacerbated by travel. Most tourists feel compelled to put the camera between themselves and whatever is remarkable that they encounter. Unsure of other responses, they take a picture" (9–10). As this quote suggests, a generic tourist would appreciate the sights in front of them, which expresses a sense of wonder. At the same time, the foreignness of what is photographed creates a sense of insecurity that can more easily be comprehended because of the distancing that is created when the camera is physically placed between oneself and what is being perceived. According to Sontag, the camera thus aids the tourist in claiming space for themselves in unfamiliar environs, just as it frames the people who are photographed as objects that are utterly alien. Understood in this way, photography is a tool of exoticization as much as painting was for Paul Gauguin. Appropriately enough, for instance Bénédicte Ledent reads Lucy's street photography of Americans scurrying about as similar to Gauguin's renditions of Polynesian people (63).

Susan Sontag moreover determines this distancing effect as one that is generally inherent to acts of photography, i. e., as not just occurring in tourist situations; or rather, photography generally lets one become a tourist in another's reality (Sontag 57). One does not need to travel far to find foreignness, as Sontag points out with regard to (newly) American photographers within the United States, who, when "[f]aced with the awesome spread of alienness of a newly settled continent, people wielded cameras as a way of taking possession of the places they visited" (65).[158] Lucy's acts of photographing thus repeat both the exoticizing, dis-

---

Noa, Gauguin writes: "A profound sadness took possession of me. The dream which had brought me to Tahiti was brutally disappointed by the actuality. It was the Tahiti of former times which I loved. That of the present filled me with horror" (*Noa Noa* (1919) 18). However, the Tahiti Gauguin was looking for in all likelihood never existed: since the island's "discovery" in 1786, the term "Tahiti" had come to transport island imaginations of paradise on earth and utopias of societies without moral corruptions. Tahiti's realities were disregarded so that in the European imagination the island became more of a mythical space (Heermann 147).

158 Sontag also points to photographers such as Jacob Riis or those of the Farm Security Administration, who depicted some Americans as foreign on the basis of class differences.

tancing gaze of the tourist turned (neo)colonizer and the possessive gaze of the American making a home. According to Sontag, this double bind of appropriation and distancing is at the heart of any photographic act; it is "one of the principal devices for experiencing something, for giving an appearance of participation" (10), even as "[p]hotographing is essentially an act of non-intervention" (11). Ultimately, "[b]ringing the exotic near, rendering the familiar and homely exotic, photographs make the entire world available as an object of appraisal" (Sontag 110).[159]

In my reading of Lucy's visual practices, the act of taking photographs with its doubled effect of appropriation and distancing, as Sontag describes it, mirrors Lucy's general ways of relating to the world and her attempts to find a place for herself in it. As I pointed out above, since she came to the United States, Lucy is continually caught in a feeling of being at sea, torn between a desire to distance herself from her former home and feeling homesick, and between the desire to adopt her new country and feeling alienated by it. Photography, I argue, with its similarly torn characteristics of approximation and removal, structurally represents Lucy's state of in-betweenness. Keeping her distance and taking possession via photography exposes Lucy's general struggles of self-making in the United States in a fashion similar to Sontag's appraisal that "picture-taking is both a limitless technique for appropriating the objective world and an unavoidably solipsistic expression of the singular self" (Sontag 122). By finding ways to relate to her environs photographically, Lucy also finds ways to express herself, which, as was seen above, is not marked by stability but rather its opposite.

Framing Others by her gaze, symbolically taking possession of them, and thus repeating colonizing gestures may provide an outlet to Lucy's desires to turn the tables and to liberate herself from the position of the colonized "island girl," but it does not relieve her of her search for wholeness. This can, for instance, be read in Lucy's handling of the photographs she has taken, when she notes about the developing process: "I would try and try to make a print that made more beautiful the thing I thought I had seen, that would reveal to me some of the things I had not seen, but I did not succeed" (160). Lucy's attempts to find hidden meanings that might reveal themselves through images recalls her flâneur-like attempts to read the faces of men by gazing at them.[160] Disillusioned by her recognition that her visual expectations of the United States were immediately ruined when she ar-

159 Maria Helena Lima describes precisely this process when she assesses that Lucy "defamiliarizes the ordinary experiences of life, making the strange familiar, and the familiar, strange" ("Imaginary Homelands" 860).

160 Moreover, the search of a hidden reality that might reveal itself through photographs takes up American photography discourses, such as when Jacob Riis, for instance, looked for the unknown reality of "the other half" (see here Sontag 55).

rived and first saw the country, Lucy here might be looking for a new set of signs for her to decode in the photographs she has taken. It even stands to reason that Lucy's manipulations of the photographs are attempts to create signs that are readable and desirable to Lucy. Although Lucy can only attempt to produce purportedly concealed meanings by manipulating the developing process of the photographs, her actions nevertheless signal a desire for the ability to create.

In conclusion, Lucy's acts of self-expression, produced for herself, highlight that artistic expression may work to constitute a sense of self, however fraught it might be – or perhaps even more: that artistic expression is able to enunciate the very precariousness of the self. "I understood that I was inventing myself, and that I was doing this more in the way of a painter than in the way of a scientist. I could not count on precision or calculation; I could only count on intuition. I did not have anything exactly in mind, but when the picture was complete I would know," Lucy realizes about her processes of individuation (134). While the present Lucy creates is not be clear cut one, through photography she is able to define it as her own, as something on which she may work as she wishes. She thus posits herself as a person with the agency and the power of creation, which gestures towards a self-made future.

At the very end of the novel, Lucy then turns from photography to writing and takes up a notebook to write her life, which suggests that she may have exhausted the possibilities of visual media for her project of discovering herself. Throughout Kincaid's oeuvre, however, visuality is increasingly made productive to engage history and questions of representation, particularly through interplay with the written word, as will be explored in the following chapters.

3.2 A Photo Album of History: Ekphrasis in *My Garden (Book):*

Jamaica Kincaid's collection of nonfiction essays *My Garden (Book):* seems invested in describing the author's garden and her activities in it throughout the year.[161] The texts appear wholly concerned with Kincaid's passion to tend to the plot of land behind her house in North Bennington, Vermont. The book opens with Kin-

161 An earlier version of this chapter was published in *America After Nature. Democracy, Culture, Environment* (Universitätsverlag Winter, 2016). I owe thanks to Verena Laschinger for drawing my attention to the link of nature and visuality in Kincaid's works. This chapter benefitted greatly from our discussions and her giving me the opportunity to present my thoughts on the panel "'Shock and awe, beauty and despair – it's all there': Eco-Photography's Views on the Transnational Landscape" at the conference "America After Nature" (Würzburg 2014).

caid's discovery of her interest in plants and her learning about bloomers, shrubs, and trees that take to the South Vermont climate, and it concludes with a travel account of her seed gathering in China to supplement her own garden with species from far away. In this, Kincaid's account of the garden and her hobby are of a rather personal nature. However, the discovery of her interest in gardening also leads to the discovery of various meanings that underlie the concept of the garden in the Western world. On one level, then, *My Garden (Book):* is merely a gardening memoir, but it also traces the gardener's realization of her own participation in colonial practices. In doing so, the book positions gardens as representations of colonial history in prompting their owners to repeat and replicate acts of colonization. *My Garden (Book):* questions the colonial history of single plants as well as the practices of transplantation and naming as they occur in gardening and botany.

While the garden is ostensibly at the center of the book, its continuous links to colonial history also bring questions of representation to the fore. Moreover, the garden itself is represented by another medium – the book. That various forms of representation really are at the heart of *My Garden (Book):*'s concerns is already hinted at by the book's title: the parentheses and the colon point to the inseparability of the garden and the book as its representation.[162] One could read the title as "my garden, which is this book, will begin after the colon," i. e., the garden is located *within* the book. Text and visual media are also intertwined in *My Garden (Book):*, as it is replete with images and a page design of leafy twines. Most interesting, however, is the occurrence of the visual *within* the verbal in this book, when images of the garden are created through extended descriptions that halt the narrative flow of the text to attend to the purported sights.

This chapter explores the ekphrastic writing in *My Garden (Book):* and its participation in negotiations of history. I argue that the complex relationship between visuality and text turns *My Garden (Book):* into a photo album of sorts that collects textual images of Kincaid's garden. Moreover, this garden is not only the space of personal pleasure and interests, but it becomes a representation of colonial history. In *My Garden (Book):*, the history of every plant is intertwined with the history of colonialism, and every act of gardening is reminiscent of colonization. Finally, my analysis of the doublings of representations of history in the garden and the book through text and images demonstrates how *My Garden (Book):* reflects on

162 Other scholars have analyzed Kincaid's critique of colonialism in her garden writing in detail. See, for instance, the work of Rachel Azima, Jeanne C. Ewert, Wendy Knepper, Melanie A. Murray, and Susie O'Brien. These scholars, however, focus more on the content of *My Garden (Book):*, while I seek to add to the discussion with an analysis of representation through visual and verbal media and their interplay in the book.

the nature of representations in general, identifies them as colonial legacies, and calls for an awareness of colonial remnants in any act of naming or framing – whether in the garden or in art.

An Ekphrastic Photo Album: Citing the Sight of the Garden

The garden in *My Garden (Book):* is Kincaid's own. The plot of land located behind her home in Vermont is large enough to have a section called "woodland" (*MGB* 16) and a "soft fruit garden" (*MGB* 18). It houses both rosebushes and a pumpkin patch. Yet the kitchen garden is not intended to feed the family; rather, it is a luxury. Like her other plantings, Kincaid grows the fruit and vegetables for her enjoyment as these are "things that would be much cheaper to buy at the store" (*MGB* 123).

Foremost, the garden exists to bring pleasure to the gardener. Kincaid describes how she tends to it, frets over it, plans next year's garden, and spends excessive amounts of money on it. It is a project that takes up time and energy all through the year. However, the most frequently performed action in the garden is looking at it. Throughout Kincaid's occupation with the garden, "watching," "getting a glimpse," "staring," "seeing," "observing" it (all verbs are found on two opposite pages of *MGB* 52–53) are at the center of her account of it.

Kincaid also compares the garden to other visual arts, such as painting or sculpture. She judges the garden to be "the most useless of creations, […] it won't accrue value as time goes on" (*MGB* 111). Yet establishing the garden as a piece of art that exists through its creator's "act[s] of will" (*MGB* 111) once more emphasizes its purpose as a visual joy, similar to paintings or sculpture. With this emphasis, *My Garden (Book):* is less concerned with the garden itself than with the visual perception of it.

Visuality already plays a role in this individual garden's history. The previous owner of the house and garden in North Bennington, Vermont, Robert Woodworth, was a professor of biology and botany at Bennington College and a member of the science faculty from 1935 to 1989.[163] The garden's former owner likely had the same (if not greater) botanical interest in plant species as Kincaid. Woodworth also was a pioneer of time-lapse photography.[164] "I do not know if the exciting and unusual

163 See the articles on Robert Woodworth in *The New York Times* and on the *Our Bennington* website.

164 According to Kincaid, he even "invented time-lapse photography" (*MGB* 31). However, his work was preceded by stopping motion in photography, which was accomplished as early as in the 1870s, e.g., by Eadweard Muybridge (see, for instance, Orvell 68–69).

collection of trilliums, jack-in-the-pulpit, squirrel corn, Solomon's seal, and mayapple that are in the bed just outside the kitchen window are the very same ones that are in his films on time-lapse photography," Kincaid writes (*MGB* 31), but her musing about it already suggests such a connection. Time-lapse photography collapses a period of time while focusing on one object. And just as Woodworth conflated time while perhaps observing the plants in his own (later Kincaid's) garden, so does Kincaid collapse time in her writing about the garden. In that, Robert Woodworth is positioned as a predecessor to Kincaid's own project in *My Garden (Book):*. While she takes up the medium of text instead of photography, in analogy to the contraction of time in Woodworth's work, Kincaid's descriptions of looking at the garden halt time in narrative progression. Static and purely descriptive passages culminate in lists of what the gardener sees (*MGB*, e.g., 27, 69, 125, 173–174). Kincaid here uses ekphrases in the classical sense to present the garden. In Hellenistic rhetoric,

> the ekphrasis as an extended description was called upon to intrude upon the flow of discourse, and, for its duration [...] to rivet our attention upon the visual object to be described. [...] It was, then, a device intended to interrupt the temporality of discourse, to freeze it during its indulgence in spatial exploration. (Krieger 7)

While the halting of narrative also freezes time in ekphrastic passages, in *My Garden (Book):* temporalities are also converged in the study of sights of the garden, as for instance in the following passage:

> (again) that wisteria blooming now (or then) so close to the buddleia which in turn is not too far from the *Phlox paniculata* "Nor[a]h Leigh," which is also somehow in the middle of the *Phlox paniculata* "David," is all pleasing to my eye, as I was looking at it then (now); at that moment of the wisteria, turning left or right (counterclockwise or clockwise), this is what I could see in front of me [...]: the perennial pea (*Lathyrus latifolius*) in bloom in its guzzling way [...], some cultivars of *Lobelia siphilitica* (bought from Dan Hinkley because I was so taken by his description, and I remain open to seeing this lobelia just the way Dan described it) on the verge of blooming, an accidental planting combination of *Platycodon grandiflorus* blue and pink (*MGB* 22).

The contraction of "now" and "then" transports a form of "seeing" the plants into different temporalities: "now" refers to the author's present while observing the garden; this is then recapitulated in the author's memory while writing about it; and finally, "now" also refers to the present of the reader, who follows the narrator's description of "looking at it then (now)." The present of the reader at the moment of their comprehension, that of the author while remembering the sight of the garden and simultaneously producing the text, as well as the gardener's present when observing the garden in front of her, thus accumulate in the word "now." The same word refers to a plurality of signifieds and thus converges temporalities. *My Garden (Book):* here makes use of repetition in the same way as the later novel *See Now Then* does in bringing forth simultaneity.[165] This conflation of temporal levels in *My Garden (Book):* equates the physical sight of the garden with its memory as well as with the imagined image of it.

Temporal entanglements purport to affect the representative capabilities of the visual. Now the visual can occur anywhere and is not tied to actual sight. *My Garden (Book):* thus suggests that the sight of the garden can be conveyed by verbal means when it proposes that text induces imaginative visions of the garden. This is also apparent in Kincaid's evaluation of descriptive text in garden catalogs: "The best catalogues [...] will not have any pictures" (*MGB* 62), as the absence of the actual image allows for imaginations as experiences of sight: "and my imagination takes over as I look out at the garden, which is a blanket of white [with snow], and see it filled with the things described in the catalogue I am reading" (*MGB* 88).

In the conflation of imagination, memory, and sight *My Garden (Book):* operates in the way of a photo album: it collects snapshots that were taken in the past. This is also evident in the book's design. Corresponding with its main subject, *My Garden (Book):*'s pivotal hue is green, which extends from the dust jacket and binding to the fly-title and title page to all printed watercolors to the design of the text pages. The title page is adorned with a centrally positioned watercolor below the book's title. Wholly in green tones, the picture shows a female figure from behind. She is dressed in overalls, a simple white shirt, and a sun hat. In a quintessential gardener's outfit, she sits by a window and looks out at a garden. The image is framed by a square, dark-green border, but its four corners are set apart in a lighter tone of green, mimicking photo corners. Imitating a photo album with this detail, the title page's illustration sets the tone of the book: in a photograph that is emulated by a drawing and watercolors, a gardener is looking at the garden.

The design of the photo corners recurs in the page design of *My Garden (Book):*. The white pages are encased by light-green frames which are decorated with trims of single strands of stylized leaves. Most noteworthy, though, is the form of the green frames that encompass the text: they are not fully square, but have little, round indentations towards the white in every corner, suggesting stylized photo corners. This design then frames the text like a photo album that collects the memories of the garden's sights. Moreover, the book contains a number of watercolors by illustrator Jill Fox, again all in green. Not all images, however, are directly linked to the text by showing what the text describes. Details differ considerably. It is rather as if the artist took her general cues from the text and

165 See chapter 1 on "The Perpetuity of the Present and the Presentness on the Past."

processed them with her own craft and imagination. Some of the images have no immediate links to the text at all (see, for instance, *MBG* 38). It would therefore be wrong to assume that these watercolors are visualizations *of* the text. They are rather representations of the garden in their own right.[166] Nonetheless, in terms of their mediality, they link image and text in that they both occupy the same white space within the pages' green frames. The watercolors "bleed" onto the same white pages where the text is located. In this, text and image present as equal here.

From the argument that *My Garden (Book):* imitates visuality in text through ekphrases and a photo album in design, one might conclude that it is the book's project to make the reader see through text, rather than through images. However, text (including ekphrases) can only produce the sight of letters and their arrangements on a page. Beyond that, the "seeing" must rely on imagination as "the material [to be seen] dissipates into the airiness of words" (Krieger xv). Or, as W. J. T. Mitchell puts it:

> A verbal representation cannot represent – that is, make present – its object in the same way a visual representation can. It may refer to an object, describe it, invoke it, but it can never bring its visual presence before us in the way pictures do. Words can "cite," but never "sight" their objects. (*Picture Theory* 152)

Kincaid seems to be aware of this impossibility, as although she appears to attempt an amalgamation of the visual and the verbal in *My Garden (Book):*, and to some extent also achieves this, she writes about the relationship of the two media: "In the narrative that we are in (in the Western one), the word comes before the picture; the word makes us long for a picture, the word is never enough to the thing just seen – the picture!" (*MGB* 130). While she specifies "the narrative that we are in" as that which is determined by the knowledge and values of Western culture, it simultaneously also refers to *My Garden (Book):*, since this is the text in which the phrase occurs. Further declaring that, in both Western culture and the text at hand, "the word comes before the picture" establishes a hierarchy between the verbal and the visual, but we might as well read this statement temporally: the word precedes the picture, or the picture follows the word, which suggests that the word may create a sort of picture through ekphrases. The passage immediately retracts this possibility, however, when it concludes with "the word is never

---

166 They rather randomly present different aspects of Kincaid's gardening and certainly are aesthetically pleasing. As my interest here is in visual-textual engagements with history, however, my analyses leave these watercolors aside and focus on the ekphrastic representations of the garden in *My Garden (Book):*.

enough" in comparison with the picture. The "ekphrastic hope," as Mitchell terms the moment in which "the impossibility of ekphrasis is overcome in imagination or metaphor, when we discover a 'sense' in which language can do what so many writers have wanted it to do: 'to make us see'" (*Picture Theory* 152) remains arrested, as ultimately sight through text can only be attempted, but never achieved.

From Plant to Medium: The Garden's Representation of Colonial History

Beyond the images of a personal life that could also be collected in a photo album, the garden also represents aspects of the world history of colonialism.[167] Kincaid condenses personal and historical memory in the garden in the assertion that it is "an exercise in memory: a way of remembering my own immediate past, a way of getting to a past that is my own (the Caribbean Sea) and the past as it is indirectly related to me (the conquest of Mexico and its surroundings)" (*MGB* 8). The example of the conquest of Mexico, led by Hernan Cortés from 1519 onward, here refers to European imperialism in the Americas in general. As I will show in the following, Kincaid positions the garden as a medium of memory and history. By highlighting the colonial legacies of gardening in acts of transplanting and naming, she showcases the garden as a small-scale colonized space.

W. J. T. Mitchell proposes to understand landscape as a medium: "It is a material 'means' [...] like language or paint, [...] a body of symbolic forms capable of being invoked and reshaped to express meanings and values" ("Imperial Landscape" 14). *My Garden (Book):* takes this up with regard to the smaller space of the garden or even individual plants. Kincaid became cognizant of the garden's representational faculties when she read William H. Prescott's *The History of the Conquest of Mexico* (1843): "I came upon the flower called marigold and the flower called dahlia and the flower called zinnia, and after that the garden was more to me than the garden as I used to think of it. After that the garden was also something else" (*MGB* 6). It is the consideration of the history of gardening and the plants grown in the garden that transforms it from merely a pleasurable sight into "something else," namely a representation of a history of conquest.[168]

167 See O'Brien, Tiffin, Ewert, Azima, Knepper, and Murray. These scholars also observe the connections of gardening and colonialism. I add to this discussion by explicitly understanding the garden as a medium.

168 It is not only in *My Garden (Book):* that plants are imbued with historical meaning. For example, in *Lucy* the daffodil becomes a symbol of colonial power. Upon seeing the plant for the first time after travelling to the United States, Lucy becomes enraged as to her it represents the impositions of colonial education that had her learn William Woodsworth's poem "I Wandered Lonely

Kincaid elaborates on the colonial legacy of gardening with the example of the dahlia's history: its Aztec name, *cocoxochitl*, and the recognition of its use as a medicinal plant as well as its cultivation for its beauty were utterly disregarded by the Europeans after Cortés's invasion. Kincaid polemically highlights the assumed European view that "these plants from far away, like the people far away, had no history, no names, and so they could be given names" (*MGB* 122). Today, the Aztec knowledge of the *cocoxochitl* is still superseded by the name "dahlia," after the Swedish botanist Andreas Dahl (*MGB* 118–119), by which the plant is now commonly known. The dahlia thus represents the conquest of the Aztec Empire as well as colonialists operating with disregard of knowledge, language, and culture of the people and thereby negating their worth and even their existence. The dahlia also shows how a plant is claimed as "European" through its renaming (or supposedly being named in the first place) after a European botanist.

Further examples of plants as media of history are the breadfruit and the hollyhock. The breadfruit was imported to the Caribbean from the East Indies as a low-cost provision for slaves (*MGB* 136), and Kincaid sees it as "not a food, it is a weapon" (*MGB* 137). Similarly, the hollyhock signifies the history of slavery in the United States: when in bloom, the hollyhock closely resembles the *Gossypium* (both belong to the family of the *Malvaceae*). *Gossypium* is better known as cotton, which has tormented people throughout Kincaid's own ancestral history (*MGB* 150), as well as throughout the American South in general. Helen Tiffin reads such associations of plants with history as part of Caribbean cultural heritage and discerns that "the practices of agriculture and horticulture were necessarily and variously associated by different Caribbean populations with dispossession, slavery, and servitude, exile and colonization" ("Replanted" 149).

The title of one of *My Garden (Book):*'s chapters programmatically declares: "To Name is to Possess." Kincaid admonishes colonial naming practices as "a spiritual padlock with the key irretrievably thrown away, [...] a murder, an erasing" (*MGB* 122). In contrast, she highlights acts of liberation from colonial powers through choosing one's own name: Rhodesia renamed itself Zimbabwe after independence from British colonial rule and the African American poet and author Amiri Baraka discarded his former name LeRoi Jones (*MGB* 122), "Bantuizing and Swahilizing" it to dispose of the American "slave name" and to change "into a blacker being" (Baraka 376). To be added is the name change of Elaine Potter Richardson, who in 1973 became Jamaica Kincaid so that her name would indicate

---

as a Cloud" by heart without ever having seen the flower that is not grown in the Caribbean. In this, the daffodil attests to the colonizer's claim to totality of Western (here British) art and sensibilities (*Lucy* 30).

the area of the world she is from, but also to free her from her former identity as a British colonial subject and the daughter of an overbearing mother to be able to write without the restraints of being Elaine Richardson (see Kincaid in Vorda 14–15).[169] In highlighting the significance of naming, Kincaid reveals the parallel between colonialists and botanists, since both "emptied worlds of their names; they emptied the worlds of things animal, vegetable, and mineral of their names and replaced them with names pleasing to them" (*MGB* 160).

The naming and renaming of plants, animals, lands, and peoples interrelates with transnational movements: European colonization was based on transatlantic movements from Europe to Africa and the Americas, first with the purpose of conquest, then with the intentions of exploiting peoples and lands and returning the spoils of this exploitation back to Europe. On a smaller scale, such transnational movement is repeated in the garden. In a move that reenacts the travels of European colonizers, the gardener Kincaid travels to China to gather seeds from Asian indigenous plants to cultivate them in her very own backyard (see *MGB* 188–219). Linda Lang-Peralta and Jeanne C. Ewert convincingly identify Kincaid's consumerism as participation in colonialist practices. In addition to the chapter in *My Garden (Book):*, Kincaid later wrote about her seed gathering travels in her 2005 travelogue, *Among Flowers: A Walk in the Himalaya.* Her crossing the line from colonized to colonizer (*MGB* 123) is especially apparent in her journeys to China and Nepal, as exploring the world with the purpose of collecting foreign plants and species repeats previous explorations of that kind in colonial times.[170] Kincaid spells this out by referencing Frank Kingdon-Ward's *Plant Hunting on the Edge of the World* (1930), Ernest Henry Wilson's *Plant Hunting* (1927),[171] Patrick Synge's *Mountains of the Moon: An Expedition to the Equatorial Mountains of Africa* (1938), and Reginald Farrer's *Among the Hills* (1910) (*MGB* 190). The author notes that these books are "a small part of how a journey like this, for someone like me, begins" (*MGB* 191). To be more specific, her journey begins not only with, but *in* these

---

169 Kamau Brathwaite, for instance, also renamed himself from his given name Edward. He emphasizes the importance of names "especially in a colonial or post-colonial situation where we have been named by other people and where it is therefore our responsibility to rename & redefine ourselves" (qtd. in Naylor 143). Similarly, bell hooks regards her own renaming from Gloria Jones as empowering (*Talking Back* 166).

170 See also Londa Schiebinger's book *Plants and Empire: Colonial Bioprospecting in the Atlantic World* on the history of plant-gathering voyages. Fittingly, Schiebinger calls such gatherers "biopirates" (35).

171 Kincaid here actually refers to "a book by Ernest Wilson, *Plant Hunter's Paradise,* an account of his travels looking for plants in China" (*MGB* 190). However, *Plant Hunter's Paradise* is a book by Frank Kingdon-Ward, published in 1937, about his travels through Burma and to the Tibetan border. Wilson's *Plant Hunting* (1927) is an account of his plant hunting journeys all over the world.

books (see *MGB* 190). The preposition "in" articulates not only that Kincaid's plant-hunting travels repeat those of the European botanists at the beginning of the twentieth century, but also that her writing about it is rooted in a charged tradition. This is also indicated by her chapter title "Plant Hunting in China," which overtly echoes the titles of Kingdon-Ward's and Wilson's books. The reference to the botanists' journeys once more highlights the roots of present-day gardening in this history of conquest as it is already set up in the texts from the beginning of the twentieth century. For instance, in *Plant Hunter's Paradise* Frank Kingdon-Ward refers to the continuation of conquest at home when he writes at the end of the book: "The journey was over. But only the journey. Not the exploration. This would be continued, year after year, at the flowering of the rhododendrons, in many an English garden" (328).

However, while repeating both the seed gathering journeys as well as the writing about it, Kincaid is conscious of her complicity in colonial practices. In "Plant Hunting in China" (*MGB* 188–219) as well as in *Among Flowers*, Kincaid becomes the tourist she admonished so vigorously in her early "angry and provocative book" *A Small Place* (Murray 116). Here, she exposes the tourist as an "ugly thing" to whom it never occurs "that the people who inhabit the place […] cannot stand you, that behind their closed doors they laugh at your strangeness" (*Small Place* 17). Yet the irony of her becoming a tourist herself does not escape Kincaid (*Among Flowers* 20), and while she is unable to remember the names of the Sherpas who cooked for the tourist group and schlepped tables and chairs up the Himalayan mountains for their comfort, Kincaid is adamant that her neglect does not reflect a colonizing relationship. It is solely her anxiety, unease, and unfamiliarity that prevent her from interacting properly with the Sherpas, she maintains (*Among Flowers* 26–27). Kincaid thus suggests that the nature of such encounters originates in globalization, rather than in colonization. In her seed gathering and gardening activities, Kincaid implicates herself in processes of globalization that are grounded in the very colonizing practices she criticizes, but in doing so, she exhibits an awareness for her participation in such relationships.[172] This focus on participation may also be read as upturning previous hierarchies. Gary Holcomb reads something similar at play in Lucy's travelling and notes its subversive qualities: "Recognizing that the subjectivity of travel writing operates at the center of colonial domination, Kincaid, with Lucy, assails the certitude written into who may write travel and who may not" (292). Holcomb's observation equally holds

172 Murray shows that "through the deployment of irony [Kincaid] reveals her awareness of the contradictions" (123) between her critique and her participation in globalized tourism. See Feder for a detailed analysis of Kincaid's position as a tourist in *Among Flowers*.

true for Kincaid's travel writing in *My Garden (Book):* and *Among Flowers.* Simply by travelling and writing about it, Kincaid upsets colonial literary standards.

Representation and Appropriation: The Book as a Simulacrum of the Garden

What becomes apparent in the consideration of the garden as a visual pleasure, as a work of art, and as signifying history within *My Garden (Book):* is that it exposes not only mechanisms of colonialism, but also those of representation. The garden in *My Garden (Book):* is positioned in a triad of signification and reference. First, the garden plants represent history based on the individual roles they have played in it. Second, it showcases the craft and with that the power of the gardener, which in turn, third, references the power of the colonizer to transplant and dominate. The garden as a space in which the power of mankind is exercised thus represents the history of colonialism. This conception of the garden is brought forth by a book that itself consists of a complex combination of visual and verbal representations, since it functions in the way of a photo album. In a suggested amalgamation of textual and actual gardens, the text draws attention to the representation of the visual (the ekphrastic descriptions of the garden) within the verbal, while what is to be "seen" through the ekphrases – the actual garden – already is a representation itself. Ultimately, I argue, the book questions not only history, but also modes of representation and relationships of representations and their referents.

Such doubling of representations is also at the heart of W. J. T. Mitchell's definition of ekphrasis as "the verbal representation of the visual representation" (*Picture Theory* 152). In their definitions of the literary figure, both Mitchell and James Heffernan are adamant about the verbal representing *another medium of representation* (see Heffernan 300).[173] In the case of the garden and its plants, which are imbued with history, the representation is not bestowed on the object by the artist, but the meaning is presented as inherent to the raw material and the sheer act of working with it, i. e., the plants themselves and their transplantations which signify on the history of colonialism. Mary Lou Emery interprets ekphrases as gateways to the "real" in that "ekphrasis seems to grant us access to that which lies beyond representation" ("Refiguring" 263). In *My Garden (Book):*, I argue, ekphrases are employed to very different ends: through the uses of a variety of representations, *My Garden (Book):* showcases the acts of representations

173 In this, Mitchell and Heffernan go beyond the definition of the rhetorical device as simply halting narrative flow through extensive description, as quoted at the outset of this chapter.

themselves, which in the context of discussions of colonial history invites us to think about possibilities of representation in general.

My Garden (Book): repeatedly references other visual artworks concerned with the garden. Kincaid compares the sight of her own garden to that of another gardener and artist: gazing at some nasturtiums growing over a walkway in her garden, she recalls a photograph of nasturtiums that lean into a walkway in Claude Monet's famous garden at Giverny. Interestingly, the representation of the flowers in Monet's garden is again doubled by the note that the image was found in another book on gardening (Wayne Winterrowd's *Annuals for Connoisseurs, MGB* 54). Comparing Monet's planting of nasturtiums to her own, Kincaid notes: "At first I felt wonderful that I had had the same idea as a great gardener, and then, unable to help myself, I felt envy, because his nasturtiums had turned out much better than mine. His looked like a painting – the way all natural beauty looks. Mine were just a planting of nasturtiums" (*MGB* 54). According to this quote, all natural beauty looks as if it were created by an artist, or, the other way around: artists of visual media are able to create natural beauty. Of course, this beauty is not natural at all but artificial. What is beautiful then, is the static representation of the natural in a visual medium, and not the organic and natural itself. Hence Monet's nasturtiums are not more beautiful than Kincaid's, but indeed their portrayal in a photograph is. The image thus conveys more meaning (beauty and naturalness) than its referent could.

Moreover, a referent may also take on the cultural meanings of its representation and in the process lose its naturalness. The English landscape, for instance, adopts the qualities of its own representation: Kincaid marvels at the English people who "obsessively order and shape their landscape to such a degree that it, the English landscape, looks like a painting (tamed, framed, captured, kind, decent, good, pretty)" (*MGB* 132). Here, the English landscape in Kincaid's writing functions as a "medium of cultural expression" in the sense of W. J. T. Mitchell's definition ("Imperial Landscape" 14). Mitchell finds a doubled representation in landscape painting: the visual art represents "something that is already a representation in its own right" ("Imperial Landscape" 14). In *My Garden (Book):* the English landscape is ordered according to ideals (*MGB* 110) and demonstrates how the imperial project not only moved outward, creating such spaces as the Botanical Garden in St. John's, Antigua,[174] but also moved "inward toward a reshaping and re-presen-

174 An exceptional space for the display of taking possession of things and people in the way of the colonizer-gardener are the public gardens of the empire. Kincaid describes botanical gardens as "the back yard of someone else, someone far away, someone's landscape the botanical garden can make an object" (*MGB* 148). In this light, the botanical garden in St. John's in Antigua, for instance, comes to represent domination, as it is associated with "the English people, their love, their

tation of the native land" (Mitchell, "Imperial Landscape" 17) as pleasant and benign.

In another examination of a painting, Monet's work once more serves to exemplify the relationship of visual art and its referents. Visiting Claude Monet's garden in France, Kincaid compares the actual garden in front of her to the paintings depicting it: while contemplating the plants before her, she "had their counterparts in Monet's paintings in [her] mind" (*MGB* 126). The image already known supersedes the actual sight. Looking at the pond's famous water lilies lying on their sides, Kincaid writes, "but on seeing them that way I immediately put them back in the arrangement I am most familiar with them in the paintings," with them sitting upright on the water (*MGB* 127). Accordingly, the perception of reality is preconceived by a sight that was "learned." Kincaid's question about whether the water lily garden would be the same without the paintings (*MGB* 126) is thus straight to the point. Especially the absence of the Hoschedé sisters, whom Monet painted in a boat on the pond, points to this interweaving of the known image and simultaneous visual perception of the same sight, as Kincaid is almost startled by missing the girls on the pond: "On the day I saw it, the pond, the Hoschedé girls (all three of them) were not in a boat looking so real that when they were seen in that particular painting (*The Boat at Giverny*) they would then define reality" (*MGB* 127). Again, the painting – the visual representation of the pond – defines what is real and not the actual sight itself. As with the photograph of the nasturtiums that defined natural beauty, here the painting defines reality. These examples of visual representations and their referents reveal the power of the likenesses over the original, as they are more "natural" and beautiful and even more "real." They even affect the perception of reality in that the images known from pictorial media superimpose themselves over actual scenes of reality.

In a last observation of visual art, a collection of glass sculptures at the Harvard Botanical Museum, Kincaid even more specifically speaks to the effect of representations. The museum displays specimens of fruit and flowers made of blown glass:

> These fruits and flowers [...] are all beautiful, and, as is the way of likenesses, seem more representative of the real than do the things that they are meant to resemble. The creation of these simulacra is also an almost defiant assertion of will: it is man vying with nature herself. To see these things is to be reminded of how barefaced the notions of captivity and control used to be, because the very fabrication of these objects [...] attests to a will that must have

---

need to isolate, name, objectify, possess various parts, people and things in the world" (*MGB* 143). Accordingly, the garden, and the botanical garden in particular, is more than a collection of plants. It is also the power of domination represented by such a collection.

> felt itself impervious to submission. How permanent everything must feel when the world is going your way! (*MGB* 79–80)

Again, the representations here hold more meaning than their referents. Furthermore, the assessment of likenesses in visual art establishes a parallel to gardening and to colonization. The creation of art is indicated as an "act of will," just like the creation of the garden (see *MGB* 111). Considering that representations are capable of altering the perception of nature, they are, on this basis, also able to alter nature itself, as the English culture landscape makes evident. In her analysis of ekphrasis in Caribbean literature, Emery contends that a verbal representation of a visual representation "reframes" its content in a way that addresses the relationships of possession the paintings represent (*Modernism* 183). Although Emery focuses on ekphrases of paintings that frame women and people of color, her assessment equally applies to Kincaid's multiple representations of the garden: through the use of ekphrases and considerations of other art works, Kincaid highlights modes of representation and thus showcases the acts of naming, framing, and transplantation in the garden as well as in art.

My Garden (Book): is a work comprised of representations and likenesses in text itself; in text that alludes to visuality; in design that suggests a photo album; in illustrations that supplement descriptive text; and not least in its topic – the garden – itself, established as a representation of colonial practices and history. Thus framing the garden in a multitude of representations, *My Garden (Book):* as a whole becomes a likeness of the garden. As the glass sculptures captivate and control their referents, so does the book. Hence it can be read as a simulacrum of the garden. It similarly takes power over its referent: although it is marketed as a nonfiction work, the average reader would not be able to verify Kincaid's account of her own garden. It may as well be a product of her imagination and writerly craft. Neither is it relevant whether the garden truly exists as described in *My Garden (Book):*. The book in this way contains the garden; it is not just an account of it. *My Garden (Book):* is a likeness of the garden even in structure in that it demonstrates the ambivalence of beauty and horror similar to the garden itself: looking at the garden, we are presented with its beauty, but beneath the surface we find that the creation of the garden is based on the principle of radically asserting one's will and of taking possession of "nature." In parallel, reading *My Garden (Book):*, with its artful design and poetic language, may be enjoyable, but when considering the form of representation that is brought forth in an interweaving of the visual and the verbal, it becomes apparent that the book appropriates nature in the same way as the garden. The multiple representations in *My Garden (Book):* thus reveal colonialist practices in the garden, but demonstrate that artistic power is equally appropriative. Thinking about the mechanisms of historiography and History with a capital *H*, Kincaid's formal highlighting of the power of representations points to the possibility of totality in any representation. Considering the formal aspects of *My Garden (Book):* thus turns it from the musings of a solitary gardener into a discussion of the workings and functions of media in general.

However, *My Garden (Book):* does not blindly repeat colonizing mechanisms. Like Kincaid, who in her garden and travel participates in global and neocolonial moves, but shows an awareness of this, *My Garden (Book):* repeats with a difference: Kincaid delights in ordering the garden according to her whims and pleasures (see, for instance, *MGB* 120–121) and in thus taking a colonizing position in it. But her joy also, and perhaps even more so, lies in the garden's own will. "The spontaneity of life in the garden defies efforts to conquer space," Wendy Knepper writes (44). The unpredictability agitates and vexes the gardener, but she also finds happiness in this excitement, and delight in the garden's ability to defy her imagination through organic growth (*MGB* 14). Gardening then is not only pleasurable because of an exercise of power, but equally because of an unspoken hope that her power might be defied by the colonized, namely the garden.

In the same sense, *My Garden (Book):* is not a static text. Knepper reads the book with Wilson Harris's notion of the living text (see Knepper 42). Corresponding to Mitchell's idea of landscape as a "medium of cultural expression," Harris regards landscape as a "living text." In accordance with the concept of a landscape-text, Knepper interprets *My Garden (Book):* as another kind of living text in that it is a space that allows for a negotiation of changing spatial praxes of home and world, just as the garden does. Moreover, the multiple representations framing the garden exhibit the framing itself. In this, colonization is not only repeated, but the repetition stages the mechanics of colonization. This exhibition of underlying processes presents an awareness of them, which Kincaid calls for: "I do not mind the glasshouse; I do not mind the botanical garden. [...] I only mind the absence of this admission, this contradiction: perhaps every good thing that stands before us comes at a great cost to someone else" (*MGB* 152). As is the book's program, *My Garden (Book):* demonstrates this doubling of presentation and deconstruction not only in content, but also through form. Hence, the standstills and the colonizing freezing in permanence that it performs through ekphrases are concurrently destabilized. The impossibility of ekphrasis does not let visual images emerge, but their evocations result in a constant back and forth between the verbal and the visual, which dismantles the static framing that they seem to establish. *My Garden (Book):* accordingly undermines its own colonizing acts: the organic aspect of the garden provides space for growing, and so does the book in showcasing its ostensible taking power of the referent through ekphrases that do not freeze the garden in permanence after all.

3.3 Photographic Product(ion)s of History: The Portraits of *The Autobiography of My Mother* and *Mr. Potter*

Four of Kincaid's monographs and shorter works feature actual photographic images: the two novels *The Autobiography of My Mother* and *Mr. Potter*, the nonfictional book *My Brother*,[175] and the uncollected short essay "Biography of a Dress" accommodate photographic portraits on their book jackets or throughout the text. As the texts' titles suggest, all of them are concerned with the genres of life writing.[176] This chapter demonstrates how Kincaid's works draw on claims of authenticity, as attributed to photography and the confessional mode of life writing in the creation of fictional knowledge about the past. In chapter 3.1 I showed how Annie John's and Lucy's engagements with photography allow them to negotiate their social and historical positions by manipulating or producing representations of themselves and others. The exploration of spaces of in-betweenness and the redirection of meanings between assumedly stable positions – be it social binaries or the components of a photographic sign – changes hitherto fixed meanings and opens ups spaces for new significations. Similarly, *My Garden (Book):*'s ekphrases explore such representative capacities of signs and media (see chapter 3.2). The entanglement of the visual and the verbal is equally productive in the medial interplay of actual photographs and fictional texts in Kincaid's later novels *The Autobiography of My Mother* and *Mr. Potter*, as the following chapter will show.

Drawing on the assumed quality of photographs to authentically represent their referents, combined with the particular positions of the images, Kincaid's works highlight the importance of presuppositions of the ontology of signs and media in the productions of (versions of) history. While in the earlier works the discovery of new meanings constituted by the images was the task of the novels' protagonists, in *The Autobiography of My Mother* and *Mr. Potter* this is transferred to the reader, who is presented with photographs directly. The photographs in Kincaid's works are few and arranged in such conventional positions that they could well be presumed illustrations to the text. However, I contend that they should not

175 I am here referring to and analyzing the books' first editions published by Farrar, Straus, and Giroux. Although the same photographs frequently appear on the covers of later as well as translated editions, they often are cropped or scaled down, and the book titles and author name are arranged differently in relation to the images. Kincaid's *Talk Stories*, too, a collection of essays that previously were published in *The New Yorker*, features a photograph of the author on its front cover, but its lack of links to the text means it is not of interest to this inquiry.

176 See chapter 2.3, "Writing between and across Genres," for analyses of the texts' play with notions of (auto)biography in "Biography of a Dress" and *The Autobiography of My Mother*.

merely be treated as complements to the verbal discourse, but as art in their own right. In *The Autobiography of My Mother* and *Mr. Potter* (less so in *My Brother* and the "Biography of a Dress") photographs are presented as products of history, but their links to the fictional texts manifest new meanings through which the photographs participate in the production of history through visual means.

Photography and Authenticity

Photographs have long been – and still are – regarded as a medium that provides evidence of past events. André Bazin notes: "The objective nature of photography confers on it a quality of credibility absent from all other picture-making. In spite of any objections our critical spirit may offer, we are forced to accept as real the existence of the object reproduced, actually *re*-presented, set before us, that is to say, in time and space" (13–14). Similarly, Roland Barthes emphasizes photography's claim to register reality, when he notes a convergence of reality and of the past in photography. When Barthes maintains that the thing or person photographed must have existed and "been there" for the photograph of it to have come into being, he establishes the image as proof of a past reality (*Camera Lucida* 76). Susan Sontag also points to this documentary quality commonly attributed to photographs: They "furnish evidence. Something we hear about, no doubt, seems proven when we're shown a photograph of it. [...] A photograph passes for incontrovertible proof that a given thing happened" (5). Based on the development of new technologies that allow digital manipulation of photographs, nowadays the reality of a photograph's referent is easily questioned, which leads Nicholas Mirzoeff to proclaim that today "photography is dead" because it has lost its claim to reality (65). Umberto Eco clarifies that the manipulation of photographs was possible long before digital technology: "We know that, through staging, optical tricks, emulsion, solarization, and the like, someone could have produced the image of something that did not exist" (223). Nevertheless, the assumption that a photograph reveals evidence of a past reality still clings to the medium, especially to photographs that (presumably) were produced before the digital age. Appropriately enough, Jens Ruchatz calls photography a "storage medium" (368) that allows us to access our memory and in this way ostensibly also to the past. Harvey Young, in his analysis of the well-known daguerreotypes taken in 1850 of Alfred, Fassena, Jack, Jem, and Renty, five Africans who survived the Middle Passage and were enslaved in the United States, as well as Delia and Drana, American-born daughters to Renty and Jack respectively, muses that "the photograph, as a historical document, reveals to the viewer, in the present, a past presence that is now absent" (57). Similar to Barthes, Young expects that the photographs' freezing of moments in time would grant him "a privileged glimpse into a moment that no longer exists" (57) and that the images of the African captives would provide him, their present-day reader, with access to their experiences in the nineteenth century (see Young 29). Likewise, Laura Wexler turns to a photograph of a young Black woman holding a white baby with the anticipation "that close, attentive readings of historical photographs can restore voice and context to historical knowledge that may have been hidden or repressed" (161).[177] These scholarly examinations of photographic images – all of them portraits – set out to salvage the photographs' subjects from history. They perform what Walter Benjamin described as the beholder feeling "the irresistible urge to search such a picture for a tiny spark of contingency, of the here and now, with which reality has (so to speak) seared the subject, to find the inconspicuous spot where in the immediacy of that long-forgotten moment the future nests so eloquently that we, looking back, may rediscover it" ("Photography" 510). Wexler's and Young's works are attempts to recover the individuals, their personal circumstances, and the circumstances of the images taken. It implies that the photographs might "reveal" such information about their subjects, if only one found the right way to read them.

The portraits in Kincaid's works are of particular interest to my investigation of the production of presences in the face of the void of History, because of photography's unique relevance to people whose history was lost. bell hooks points out that "[w]hen the psychohistory of a people is marked by ongoing loss, when entire histories are denied, hidden, erased, [photographic] documentation may become an obsession" ("Photography" 48). David Boxer, collector of Jamaican historic photographs, exemplifies this when he explains his fascination: "This colliding of time, the familiar seen through the veil of history, is precisely the quality in early Jamaican photography that made me want more and more. […] each work I acquired […] offered a window back into the past and allowed me to relive in a way that past which had been the present of my parents' parents and the present of generations of my ancestors even back to the generations just emancipated from their centuries of bondage" (8–9). Photography then becomes a central medium to "keep history" (hooks, "Photography" 48). Harvey Young, as well, equates family photographs with "history" itself (56). Moreover, photography may not only provide access to a past of which representations are scarce, but it equally plays a role in the recovery of a subject position in the present, when the camera here becomes a tool for positive self-representation to counter racist representations of "the Other" (see, for instance, Willis, hooks, Young). Photography thus has a double

177 A comparable position is held by Douglas Daniels, who asserts that "[s]cholars can use photographs to study historically 'inarticulate' segments of our population" (9).

function here: to record the past and to resist racist misrepresentations. Accordingly, the power of evidence attributed to photography is amplified by these social and historiographic functions.

I am here concerned with the specific form of photographic portraits of people of color in the transatlantic context that (presumably) were taken for private purposes and in the past, i. e., before digital manipulation became common practice. This photographic genre implies a set of yet more specific attributions to the images, since personal portraits and family photographs perform distinct functions in engagements with history because of the likely familiarity of the depicted. Siegfried Kracauer distinguishes here between contemporary and "old" photographs: "[P]hotography, which portrays phenomena familiar to contemporary consciousness, provides access of a limited sort, to the life of the original," he writes (429). The contemporary photograph allows its beholder to immediately recognize the photographic subject. They do not (have to) focus on details, but see the subject for themselves at first glance. Such a photograph accordingly performs the mediating function of an optical sign for the depicted (Kracauer 429). However, according to Kracauer, it is only within the parameters of space and time that a photograph can record what it depicts (425). The content or meaning of what has been photographed are contained, in contrast, in what he calls the "memory-image," (426) which is not linked to dates or temporal distance as a photograph is. Contemporary photographs can accordingly provide access to what they portrays only in the sense that they may evoke a memory-image of it. In her discussion of different approaches to personal photographs, Patricia Holland terms the processes of accessing an image's content as "using" it if the onlooker already knows the world depicted in a photograph. A person unfamiliar with the photograph and its subject would perform a "reading" of it. According to Holland, users have intimate connections with the photographs: the images were taken of or for users, or they were inherited together with a surrounding knowledge (121). Holland's users and Kracauer's beholders of contemporary photographs are able to access information about what an image shows without excessively studying it. Their reading, or "studium" – to use Roland Barthes's term (*Camera Lucida* 26) – exhausts itself in the recognition of an already familiar subject. Any knowledge about that subject, however, is not located within, but outside of the image; it is invoked by it, it was already available to the beholder before they looked at the photographic representation of the subject.

An "old" photograph, in contrast, calls for different attentions. Its subject cannot simply be recognized as it is not a priori known to its beholder, "the immediate reference to the original is no longer possible. [...] If one can no longer encounter the grandmother in the photograph, the image taken from the family album disintegrates into its particulars" (Kracauer 429). Once the person is not recognizable at first glance, either because of a temporal remove, or, as I would like to add, because the face has never been familiar in the first place, the focus of the beholder must shift to details in the image that might help to identify the person or thing that has been photographed. A different studium is required here to come to conclusions about both the image and what or whom it portrays. Holland aptly opposes the "user" to a "reader," to whom "a hazy snapshot or a smiling portrait from the 1950s is a mysterious text whose meanings must be teased out in an act of decoding or historical detective work" (121). For the reader, "old" photographs "lose their specific meaning in favor of more general insights into social and cultural conventions" (Ruchatz 372). This form of studium is more analytical, since no memory-image would interfere with the study of the photograph and cultural conventions are consciously taken into account of the analysis of the image to grasp its content.

Jamaica Kincaid's works present cases of purposeful blending of the personal and the unfamiliar when reproductions of seemingly private portraits are publicly printed on the covers of books that play with the notions of life writing's authenticity and are not meant for private use but to be widely sold. Hence I am here interested in the amalgamation of the two forms of studium, that which conjures up a memory-image that seemingly underlies the photographic representation but really is projected onto it by the beholder who is familiar with the image's subject, and that of the old or unfamiliar photograph that demands a more detached and detailed "reading" or examination.

Of the photographs in Jamaica Kincaid's oeuvre, I am particularly interested in those that are part of *The Autobiography of My Mother* and *Mr. Potter.* Although both *The Autobiography of My Mother* and *Mr. Potter* have been read as parts of the author's "family album" that she produces through the body of her works (see Nasta), I contend that the texts are less concerned with individual characters – and even much less with Kincaid's own parents Annie Richardson Drew and Roderick Potter – than with the collective memory of Caribbean colonial history and its still-lasting effects on present-day society and relationships. This begs the question of how photographic portraits that show individual people, and thereby document the very existence and presence of these individuals in front of the camera, may contribute to the imaginative production of knowledge about history.

A Portrait of a Fictional Protagonist: Creating Authenticity

The Autobiography of My Mother features two photographs: The sepia-toned cover image shows the reproduction of an Afro-Caribbean woman's portrait, and the backside presents an author photograph of Kincaid in black and white. To be clear from the beginning, the woman on the cover of *The Autobiography of My Mother* is decidedly not Kincaid's mother, as can easily be discerned by consulting a public family photograph such as that of Annie Drew, Jamaica Kincaid, and Kincaid's daughter Annie Shawn by photographer Mariana Cook in her collection of family portraits (21). Apart from skin color, headkerchief, and a flowered dress, Annie Drew in Cook's portrait and the woman in the cover image hardly have any physical characteristics in common. The woman on the front cover of *The Autobiography of My Mother* is young, perhaps middle-aged, with brown skin that suggests both African and European ancestry. Her dress indicates that she might be Caribbean, as she wears a variety of the douillette (or dwiyet). This is the Creole dress of the Caribbean islands that were colonized by the French, particularly Dominica and St. Lucia, which came into fashion in the eighteenth century and according to Dominican folklorist Cissie Caudeiron was frequently worn until the 1930s (see "The National Dress" and Caudeiron). The costume here includes a long flowing jupe (skirt) with a large floral design, a white chemise tucked into the jupe, slightly trimmed with embroidery and with half-length sleeves. A printed foulard (scarf) tucked into the waist of the skirt front and a madras headpiece complete the outfit. With this costume, the woman portrayed not only presents a traditional Caribbean (Creole) dress, but also what Cissie Caudeiron identifies as the direct offspring of the Sunday dress of a female enslaved or freed woman (32). The figure moreover wears a necklace of dark polished beads that is wrapped around her neck several times, and eardrops of similar but smaller beads, as well as a simple metal bangle of about one inch width on her left wrist. These adornments are part of the Creole dress (Caudeiron 33).[178]

In the portrait, the woman's lower legs and feet are not visible as the photograph (with the book cover) is cut off at about the height of her knees. Her right shoulder is slightly turned away from the camera, which results in a not quite but almost frontal pose. Her right hand is not visible, as it is hidden in the folds or a pocket of her skirt. Of her right arm only her elbow is visible between shirt and shirt. Her left hand rests on her left hip, which gives the impression of her arms akimbo with both elbows standing out. Still, her left hand is bent back, so that her arm actually rests on her wrist, which appears rather as a leisurely pose than a tense one. Her face looks relaxed, a slight hint of a smile plays on her lips. While she seems to look right into the camera, her gaze also appears distanced and disengaged, which conveys an impression of inaccessibility. Altogether, her posing conveys the assumption of this being a photograph produced

---

178 See also page 81 of Michael Ayre's picture volume *The Caribbean in Sepia* for a portrait collection of women in this traditional costume in the 1870s.

with intent and purpose, perhaps even taken at a studio. The lack of background as well as the subtle lighting from the upper left would corroborate this assumption. It can further be inferred that the woman may herself have commissioned the portrait, since she seems to be posing freely. Then again, her leisurely yet also adverse pose and the disengaged look on her face may also suggest that someone else commissioned the photograph, with which she complies, albeit half-willingly, as expressed by the small resistances in her posing. This suggests that this could be a type portrait of "la belle creole" as often taken in studio settings and collected by photographers at the end of the nineteenth century to depict the "picturesque other."[179]

The photograph moreover shows markers that suggest its age. It is entirely in sepia tones, which seem to bespeak its age, since it either yellowed over time or was produced in an age when sepia toning was still fashionable, before color photographs became popular. On its right side, water stains have damaged the image, which also suggests that it has existed for some time. However, the stains do not disturb the image; rather, their blending of lighter and darker sepia tones corresponds well with the rest of the colors and shades in the picture. It might even be speculated that the stains were placed there on purpose to evoke the impression of an authentic old photograph. Small specks of dirt in the upper half of the portrait may be interpreted along the same lines, as well as a lighter line in the bottom left corner that looks like damage from a dog-eared corner of the photograph. Another indication for possible photo retouching is the coloration of the space around and behind the female subject. The background has the same color as her blouse; only the space in her immediate proximity is darker, as if she threw a shadow. However, it is more likely these areas were darkened on purpose to create a contrast to her figure, since the darkest tones are to her immediate left, although a light source must have been positioned on that side, as can be concluded from the lighter areas in her face and the darkest space of the image being located on her other side around the area of her left hand.

All in all, neither the background nor the subject herself unequivocally reveal the purpose of the portrait, and a close inspection of both raises more questions than it provides answers. It is even possible to suppose that the image was specifically designed and produced as a cover image for *The Autobiography of My Mother*, especially considering how strikingly well the picture fits with the narrative of Xuela, who also is a woman who stands by herself and whose attitude to her circumstances and to history would very well be illustrated by her arms slightly

179 For examples see Boxer and Lucie-Smith 230–237, or Ayre 21, 70–77, esp. 79–81, 195–197, 230–233.

akimbo and with a hint of a smile on her face. However, since Xuela is a fictional character and could therefore not possibly appear in an (authentic) photograph, a search for photographic evidence, i. e., "the thing that has been there" in Barthes's terms, could only conclude that *a* woman in the dress described stood in front of a camera's lens at some point in the past. Who this woman is or was, however, remains obscure. The photograph thus provides evidence of the existence of a person, but it does not offer knowledge about that person. Meaningful insights into whom an unfamiliar photograph portrays can hence hardly be deduced from the sole consultation of that photograph. Moreover, in the photograph in question here the subject cannot be further identified, as her name is unknown. The place and time when the photograph was taken can likewise only be supposed, since the hints that the photograph provides may as well be manipulated.

Ultimately, this interpretation of the portrait must be a reading – i. e., an attempt to decode the image – solely of my own.[180] Since I have no other information about the portrayed, I can only revert to "historical detective work," as Holland puts it (121). In my reading of the image, I thus follow the claim that images "must be understood as a kind of language; instead of providing a transparent window on the world, images are now regarded as the sort of sign that presents a deceptive appearance of naturalness and transparence concealing an opaque, distorting, arbitrary mechanism of representation, a process of ideological mystification" (Mitchell, *Iconology* 8). The image nevertheless may reveal other insights to other readers. Laura Wexler astutely notes that "[p]erspectives on photographs, as on history, differ from individual to individual and from social location to social location" (161). However, one thing becomes clear: whatever meanings such a photograph may convey, they emerge from the interpretation of its beholder and are not inherent to the photograph itself.

As open as this photograph may be to interpretation, through its ties to the novel's text, meanings are both suggested and withdrawn again. As seen above, the interpretation of the photograph on the cover of *The Autobiography of My Mother* does not yield fruitful insights without verbal context. Harvey Young and Laura Wexler also rely on text, such as captions or photographers' records, to conduct their analyses, which indicates that even in such thorough analyses of historic portraits as Young's and Wexler's, the study of only the photographs themselves does not necessarily yield satisfying results and that verbal discourse about their subjects needs to be taken into account. Hence, while I agree with

180 In the only scholarly consideration of the image to date, Jana Evans Braziel indicates the photograph as "enigmatic, its reference ambiguous" ("Transmutations" 93), without commencing further investigation concerning the subject.

Young that photographs may collapse "the past with the present with the promise of extending it to the future" (50), I would challenge the assumption that they indeed may give us access to the past or restore hidden parts of history (in Wexler's terms). While Wexler argues that the historical contextualization of photographs may open a kind of wound "that offers a way for a subjugated history to be spoken" (160), in Kincaid's case, it is not historical, but contemporary contextualization that opens the photographs to new ways of engaging history.

However the photograph originated, it was consciously selected to be reproduced on the cover of a novel entitled *The Autobiography of My Mother.*[181] If the text were an autobiography, based on convention, it could safely be assumed that the book's cover shows its protagonist, i. e., the portrait of a veritable person. This cannot be the case, as *The Autobiography of My Mother* is a work of fiction. The label "a novel" at the bottom of the cover should make it sufficiently clear that the text's protagonist could not be depicted in a photograph, as Xuela is a fictional character. The term "autobiography" in the novel's title nevertheless evokes such assumptions. The impression that the woman on the cover and the novel's protagonist might be identical is corroborated by the appearance of the same image throughout the book: before the first page of text, the novel begins with a mostly blank page with the print of a small corner of the photo in its upper left corner. With the beginning of each subsequent chapter, an additional snippet of the cover image is revealed. This gradual exposition of the portrait implies an analogy to the progression of the text, in which each chapter discloses more about the life of its main character Xuela Claudette Richardson. Finally, with the last chapter, the whole picture of the woman on the cover is shown, as well as the text's reader now has presumably been presented with Xuela's whole life story.

The proximity of the image to the novel's content thus implies a connection of the woman portrayed in the photograph and the woman portrayed by the text. This suggestion of identity creates a cycle of redirection from the historic portrait that one is presented with at first glance, to at closer examination a perhaps artificially produced cover image that is unrevealing because of a lack of historical context, to a depiction of Xuela Claudette Richardson, whose life story the reader is familiar with through the novel's content.

The lettering of the book title and author name, which are printed onto the book cover and its photograph, similarly participate in processes of redirection. The title and author name are printed in two lines at the top of the cover above the woman's head and mostly on the background of the portrait, which leaves

181 Inquiries to the publishing house Farrar, Straus and Giroux concerning the selection process unfortunately produced no response.

the image's subject largely untouched. Interestingly, title and author name are not separated by space or differing typeface as they are set together in two lines in handwritten letters. They are even further tied together through the preposition "by": *The Autobiography of My Mother by Jamaica Kincaid.* Before the connecting preposition, a very slight and almost horizontal line floats between the words, which could be read as the hint of a separating dash that fell off its line. However, because of its crooked angle, it is more likely that this small line is part of the following b, which was interrupted by a momentary lack of ink. The close relation of title and author once more suggests a proximity between the author's person and the text, as already advanced by the genre marker "autobiography." Then again, the title also calls attention to the fragility of such relations, since writing the life story of one's mother would be a biography. The term "autobiography" adds the self of the author to the title, which questions the relationship of author and biographical subject. In this way, the book's title picks up on the mode of redirecting from one meaning to another.

The title and author name are moreover printed in handwriting. A comparison of the letters reveals that the two lines indeed appear to be written by hand before being reproduced on the cover, as recurring letters are not identical and show irregularities.[182] A notion of authenticity is attached to handwriting, since the assumption behind this form of script is that it must have been produced by a veritable person. Its effect is thus comparable in one sense to that of a personal photograph, since the writer, just as the photographic referent, must have existed. This effect is yet again amplified by the word "autobiography," which in itself suggests a personal and authentic written account. However, just as the cover image is not a real photograph, but merely a reproduction which possibly might even have been manipulated, this is not actual handwriting, but solely a reproduction of what appears to have been written by hand. The cover and its writing then do not present a personal, handwritten document, but solely a production that advances this impression.

That this indeed is not an autobiography or even a biographical text is made clear by the label "a novel." Distinctly not part of the title, these letters are not capitalized and are located in the lower right corner of the cover, i.e., in the farthermost location from the title. In this way, a solution to the ambiguous title is provided, but it is positioned most unobtrusively. The label "a novel" of course relieves the author of the demands of authentic life writing, which the title and

---

182 To clarify, this is in most likeliness not Kincaid's own handwriting, as already a brief comparison with her signature evinces.

its relation with the author name prompt. Nevertheless, the questions of truth, fiction, and their relationship are already proposed.

The author photograph on the back jacket additionally ties into this set of relations. A black and white portrait of Kincaid extends over the whole back of the book, bearing striking similarities to the cover image: portrayed by herself, Kincaid wears a long flowing frock that could be a dressing gown. Though more detailed, it shares a floral print with the skirt of the woman on the front. Both attires share a loosely falling material, half-length sleeves and a V-neckline. Kincaid is situated before a neutral background, which in this case is black. Further similarities are in their framings and positions: Kincaid, too, is tightly framed and the ratio of her body to the overall image is the same as that of the woman on the front. As she sits in a dark chair that is almost completely concealed by her figure, her left shoulder is slightly turned away, which gives her body an angle to the left, i. e., the opposite side as in the other portrait. She rests her head on her left hand, which is bent in a similar angle as her counterpart's left hand. Similarly, both of Kincaid's elbows are bent – though not outwardly, but close to the core of her body and her right hand is hardly visibly between the folds of her gown. Her lower legs and feet, too, are not visible, but cut off by the borders of the portrait. Kincaid's face has a serious look, and the fact that she is holding her head in her left hand gives her a thoughtful expression. Just like the woman portrayed on the cover, Kincaid looks directly into the camera, but as her head is slightly tilted to rest in her hand. She evokes the impression of dreamily – or disinterestedly – looking past the camera and, with that, past the photograph's beholder. In this portrait, the light source is located to its subject's right side, i. e., to the opposite side in comparison to the cover photograph. This image could well qualify as professionally produced for the cover of an autobiography and hence corresponds particularly well with the one on the front.[183]

When the book jacket is folded out fully, the two portraits face each other and because of their identical proportions relative to the overall photos, the nearly identical angles of their subjects' positions, and the light source we might imagine between them, they correspond exceptionally well. In this juxtaposition, author and novel, i. e., reality and fiction, are confronted with each other, just as in the arrangement of title and author name on the cover. Moreover, the similarity of the two portraits once again creates a proximity of the author and her fictional work. Between both photographs, one finds the title *The Autobiography of My Mother by Jamaica Kincaid.* If one reads the portrait on the front cover as Xuela, or at least as a representation of the novel's content, the correspondence

183 Indeed, it was also taken by Mariana Cook (1992).

of the two portraits would imply that the text, as represented by the title on the book's spine between them, was created between both of them – between the woman portrayed on the front and the one on the back. The author portrait thus also partakes in blurring the lines between the genres of (auto)biography and fiction. It creates space for speculative musings about authenticity and fiction similar to those advanced by the portrait on the front cover.

In this way, the photographs and their placements take up the play with referents that is also at stake in the novel's title. In a confusion of preconceived notions of both genre and photography, the image with its ascriptions of being an "authentic" medium depicts a fictional character. It lends its assumed credibility to the text and supports a reading of the novel as life writing. The front cover photograph participates in positioning Xuela's story as a historical biographical account. Reading this the other way around, and assuming that this indeed is an old photograph, Kincaid's novel provides a narrative to the image in the sense of Young's and Wexler's readings of historical photographs that seek to restore voices. Ultimately, we do not know whether this is an old photograph or one produced and manipulated specifically to cover Kincaid's novel, but the medial mechanisms remain the same in their suggestions of authentic representation.[184] It does not matter whether the photograph is as "fictional" as the text; in their interconnection both enact a presence where previously there was none, performing medial conventions to produce the historic account of a life to which we would otherwise have no access.

The Image of a Fictional Father: Creating a Presence

The images on the book jacket of *Mr. Potter* work to similar ends. The wrapper here features three photos in total. The inner backflap shows a small author photograph of Kincaid, though this is of less interest for the absence of direct ties to the novel. The front cover displays the portrait of a boy, standing in front of a wooden swing door, holding what could perhaps be a rolled-up newspaper. This novel, too, evokes the genres of life writing with a daughter-narrator who as the fictional author pens the life of her father, the eponymous character, to establish a relationship with an absentee parent. Following the conventions of biography, an identity of the boy on the cover of *Mr. Potter* and the novel's protagonist may be presumed. Most inter-

184 Kincaid's works here make another characteristic of photographs productive: "Photographic 'literacy' is learned. And yet, in the real world, the image itself appears 'natural' and appropriate, appears to manifest an illusory independence from the matrix of suppositions that determines its reality" (Sekula 86).

esting, however, is the third photo, the portrait on the back jacket, as it further plays on Kincaid's own family history.

The image on the back of the book shows an elderly man in a wicker chair, flanked by two women standing to his sides. The photograph is sepia-toned, which blends with the rest of the jacket design in similar beige tones. The seated man is at the center of the portrait. He is past his prime and could well be going on sixty years of age. He appears to have dressed for the photograph, wearing an impeccably white double-breasted suit, a white shirt with a classic Kent collar, a dark tie, dark shoes polished so diligently that they reflect the light, and matching dark socks. His legs are crossed, and in a leisurely pose he has folded both his hands around his upper knee. The women to his sides also are attired well, in almost matching bright dresses. The woman to the left stands slightly closer to the man in the center of the portrait. She could be about the same age as him. Her dress is white or perhaps colored in a discreet pink (which is difficult to discern because of the image's sepia tones) with a pattern of small merging circles in different colors. Both hemline and sleeves are half-length and just cover her elbows and knees. Her hair is neatly tied back and she is adorned with simple jewelry of small metal hoops in her earlobes, a necklace, and a narrow bangle on her left wrist. The third finger of her left hand shows a metal ring, which is likely a wedding band, given its placement. This hand is laid on the man's right shoulder, but not resting there in a relaxed position – her thumb and little finger are not laid down, but hover in the air just above his shoulder. On the other side, a younger woman stands upright in the same pose as the woman on the left, also facing the camera, in a dress of the same style, but of different cloth. Matching her white dress, she wears white leather shoes and a white hat. She is wearing similar earrings as the woman on the left and a watch on her left wrist. Her hand is also placed on the seated man's shoulder, and her fingers do not curl affectionately, either, but are kept straight and, in an unusual angle, do not point to the front but toward the man's head, which gives the physical touch an awkward and staged look.

The three of them are located before a blank wall, with the corner of a rug showing in the foreground of the image. The wicker chair is of a plain design, and some strands are coming loose at the bottom of the front legs. Taking all of these details into consideration, it is most likely that that this is a staged family photograph. The age difference between the women and the wedding band suggest that the woman on the left may be the seated man's wife, with the woman on the right being their daughter, although, of course, other family relations could be at play. The positioning of the man with the women to his sides nevertheless indicates a familial connection with the patriarch at the center and the women demonstrating their connection to him through the placement of their hands.

This portrait does not extend over the whole back side of the dust jacket as do the images on *The Autobiography of My Mother*, but it is integrated into a design. The edges of the images are blurred and blended with the background of the book cover. Here too, then, the two women are only half-visible, their outer shoulders fully effaced as their faces begin to merge with the background. They are thus rendered less important not only by their positions within the portrait, but also by being obliterated in being blended into the background of the jacket design. In this way, the man in their midst is positioned at the center of interest not only in the original portrait, but also in the design of the book's dust jacket.

The background of the jacket is based on the reproduction of a threadbare woven material that looks old and worn and appears to resemble the binding of an old book or journal. The worn look of the suggested binding adds to the impression of something that has traveled through time, just as the photograph has. This is further corroborated by an array of flyspecks all over the image, which in effect are similar to the watermarks and the dog-ear of the portrait on the cover of *The Autobiography of My Mother*, in that they suggest that the image may have existed or been on display for a longer period of time to have attained the smudges and damages.

That a book often read as the biographical account of Kincaid's own father Roderick Potter shows a classic private family portrait, easily leads one to assume that the image might indeed represent Roderick Potter and some of his relatives. This is further corroborated by the image's copyright note: "Back jacket photograph courtesy of the author," which establishes a connection between the author, her family, and the novel's characters. Yet the information that Kincaid provided the image neither confirms nor proves that the picture indeed was culled from Kincaid's own family album. She may as well have bought it at a flea market.

The redirection of the portrait's meanings here works similarly to that performed by the images of *The Autobiography of My Mother*. Ways to read the image on *Mr. Potter* range from private but unfamiliar family photograph to illustration of a fictional text to again private photograph shared with the reader by the author Kincaid. This redirection suggests that if the photograph depicts the book's Mr. Potter, this in turn advances a biographical reading of the novel and attributes a truthfulness to Kincaid's fictional account of a character thoroughly determined by colonial history.

It is in this intermingling of the authentic and the fictional, in both the novels' texts and photographs, that the productive moment in their engagements with the past lies. The texts provide information that convention and references lead one to read as context to the portraits. As the portraits themselves reveal rather little, their contexts must aid the production knowledge about them and about the past. While a photograph provides historical reference in that it evidences a moment in the past, without text, such as a caption or inscription, the image loses its singularity, or, in Kracauer's terms, the memory-image of what it depicts has faded. In his "Little History of Photography," Walter Benjamin aptly asks: "Won't inscription become the most important part of the photograph?"(527). Marianne Hirsch argues that "[m]ore than oral or written narratives, photographic images [...] function as ghostly revenants from an irretrievably lost past world. They enable us, in the present, not only to see and to touch the past, but also to try to *reanimate* it by undoing the finality of the photographic 'take'" (*Generation of Postmemory* 36, emphasis mine). The "fragmentariness of the two-dimensional flatness of the photographic image, moreover, makes it especially open to narrative elaboration and embroidery" (Hirsch, *Generation of Postmemory* 38). Or, as Allan Sekula puts it, "the photograph, as it stands alone, presents merely the *possibility* of meaning. Only by its embeddedness in a concrete discourse situation can the photograph yield a clear semantic outcome. Any given photograph is conceivably open to appropriation by a range of 'texts,' each new discourse generating its own set of messages" (91, emphasis in original). Without narrative, photographs reference a moment in the past, the moment when the camera's shutter closed, but it is only with difficulty that they can be placed in historical context or be attributed to individuals. In this sense, while the photographs in Kincaid's works do not lose their authenticity as historical products, they nevertheless become blanks for interpretation and reanimation in the present, which allows for the connection of fictional text and photograph.

In his work on photography and memory studies, Jens Ruchatz productively draws from C. S. Peirce's semiotic model in a way that is helpful to understand how the the meaning of the photographs in Kincaid's works is redirected. Ruchatz argues that private photographs are usually read by way of their indexicality (371), i. e., their signification of their origin. The images here do not signify the moment they were taken in spatial and temporal parameters, but rather meaning that is held by the memory-image (Kracauer).[185] In Holland's terms, this concurs with the "user" of a private photograph, who is taken to a world of memories by seeing the image and who focuses on its content. In contrast, "collective memory favors the photographs that support a symbolical reading" (Ruchatz 372), i. e., a decoding that is based on conventions and norms and, with that, a more detailed studium or reading of the image. The images in question here can hardly be read indexically,

185 The photographs' iconicity is not as pertinent here, "[a]s *icons* photos signify by similarity. The light that is reflected onto the light-sensitive emulsion delineates the objects in front of the lens in such a manner that no code is required to recognize them" (Ruchatz 371, emphasis in original). Although the unfamiliar reader would not recognize a photographed person as an individual, this person will nevertheless be identified as a human being with certain physical characteristics.

since the people depicted in the portraits would be unknown and unfamiliar to the readers of Kincaid's novels. The circumstances in the photographs were taken can only be assumed, since any text that could confirm the images' origins is (largely) absent. These photographs are not provided with defining captions, but they are linked to whole novels that suggest certain interpretations without ever confirming them directly. In this way, the images that show individuals are opened to a variety of interpretations, since none of them can be proven without a doubt. However, the fictional texts appear to relate the life stories of the portrayed, which seems to allow the readers to become users of the portraits, since now they have access to the world of those the photographs depict. The fictional narratives aid the imaginative reanimation of the portraits, as they provide a basis for (fictional) memory-images of the persons they depict. In this sense, the photographs are once again being read indexically, though now they do not signify their own origins, but the fictional contents of the novels. A shift occurs here from solely being able to discern that something or someone has been there to the assumption that certain persons (here the fictional characters of Kincaid's novels) have been in front of a camera.

It is in this redirection that the productivity of enactment of history lies, as the shift from a symbolical to a fictional indexical reading creates the potential for new knowledge to be created, which nevertheless retains the same status of authenticity as truthful accounts of the past – as, e.g., real biographies of Kincaid's parents would. In addition, this fictional indexical reading is still open to symbolization. Xuela Richardson and Roderick Potter are not individual characters, but they represent a collective affected by colonial history.[186] When the likenesses of unidentified individuals on Kincaid's book covers are opened to a variety of interpretations, the persons they portray come to represent a variety of people. Thus, it is not only that the life stories of Xuela and Mr. Potter are authenticated by their interactions with the photographs; their paradigmatic nature is also endowed with the status of authenticity. The photographs' referential power as products of history persists and, through the purported connection to the texts, works to authenticate the texts' fictional productions of history.

---

186 I argue this in detail in chapters 2.2 and 4.2.

4 Bodies: Embodied Enactments of History

The human body is the third medium considered by the present study, in addition to text and image, that Jamaica Kincaid employs in her works to engage with history. Her prose texts repeatedly foreground bodies through detailed descriptions of their attributes and functions. While I cannot consider actual physical bodies, the bodies I am interested in are mediated by text and image. Hence, my analysis of human bodies builds on the results of the previous two chapters. Nonetheless, within the visual and verbal printed media that make up Kincaid's oeuvre, bodies insist on their presence and make themselves known. As I will show, bodies occur as sites where concepts of selfhood and history are negotiated – both for the characters, who engage with their own bodies within the narrative, and for the texts, which refer to bodies as discursive sites in order to debate colonial discourses.

In these colonial contexts, history determines all notions of the human body in terms of the othering mechanisms which make it central to any colonial project, and which are imposed on bodies. Conceptions of race and gender, and of humanity itself, are negotiated through human bodies. Michael Omi and Howard Winant argue that "[s]ince the historical encounter of the hemispheres and the onset of transatlantic enslavement were the fundamental acts of race-making, since they launched a global and world-historical process of 'making up people' that constituted the modern world, race has become the *template* of both difference and inequality" (106, emphasis in original). Understood as the "template" forming structures of behavior of the powerful and the powerless, race – which is constructed as a visually perceptible category by way of so-called phenotypes – also constitutes the template for ascribing meaning to bodies. Omi and Winant further point out historic parallels in the social statuses of women and enslaved people, thereby correlating the discursive structures of race and gender: "From conquest and slavery on, racial parallels and racial 'crossings' have shaped gender relations. Women and slaves were at best lower-status humans, at worst not human at all. [...] The corporeal distinction between white men and others over whom they ruled as patriarchs and masters, then, links race to gender, and people of color to women" (Omi and Winant 107).[187] The crucial point in this for me is the comparable approach to bodies that are read as different from the white, male, able bodies that are posited as the norm. Guillermina De Ferrari even regards the body as the key component to Caribbean thought. She estimates that "the symbolic appropriation of the body by medical, legal, and political discourses is the true origin of

187 Omi and Winant here solely refer to historical structures of othering. This is not to be read as a general equation of racism and gender inequality.

 https://doi.org/10.1515/9783111027500-006

the Caribbean, a process that accounts for but is not limited to the movement of bodies and cultures that is usually the focus of Caribbean theories" (2).[188] Issues of appropriation of the body in terms of race and gender discourses are also at stake in Kincaid's works and are at the center of the following analyses. I argue that through bodily practices, such as behavior and gesture, but also by grooming and dressing their own bodies, the characters in Kincaid's texts engage with concepts of the body that are shaped by colonialism. By engaging with their own bodies and by using them to interact with others, these characters both perform and question the categories of race and gender.

Human bodies become apparent as media to contest colonial history, when the characters in Kincaid's works discover that they are not neutral beings and may not be shaped solely by self-determination and a human's particularities, but that they are always already inscribed by discourse, which is largely linked to the visually perceptible features of their bodies. Elizabeth Grosz maintains that

> bodies cannot be adequately understood as ahistorical, precultural, or natural objects in any simple way; they are not only inscribed, marked, engraved, by social pressures external to them but are the products, the direct effects, of the very social constitution of nature itself. It is not simply that the body is represented in a variety of ways according to historical, social, and cultural exigencies while it remains basically the same; these factors actively produce the body as a body of a determinate type. […] [R]epresentations and cultural inscriptions quite literally constitute bodies and help to produce them as such. (x)

In Kincaid's novel *Annie John*, the prepubescent protagonist mistakenly assumes her body as a neutral entity and the experience of being ascribed with meanings and expectations as well as the recognition of such cultural and societal inscriptions are perceived as traumatic. In this, *Annie John* represents processes of gendered inscriptions on a pubescent female body and strategies of resistance to such inscriptions, which the later novels *Lucy* and *The Autobiography of My Mother* explore more radically.

In terms of racial inscription, Hortense Spillers notes that, in historical terms, the marking of the body indeed may be understood quite literally. Referring to the Black enslaved body, Spillers discerns that the inscriptions of brutalization in terms of wounded and scarred flesh, missing limbs, or the branding of a body denoting it as another's property indexically signify not only the past injury, but also the body's abject status as a destructible entity (in opposition to a body protected by juridical and moral discourses of humanity). These physical markings on the assaulted Black body, Spillers argues, "render a kind of hieroglyphics of the

188 De Ferrari here foremost refers to Édouard Glissant's poetics of relation.

flesh whose severe disjunctures come to be hidden to the cultural seeing by skin color" (67). In other words, the body and its visible characteristics, here color of the skin, now signal its vulnerability. Such inscriptions, attributed and read by others, then are difficult to overcome, as these "markers [are] so loaded with mythical prepossession that there is no easy way for the agents buried beneath them to come clean" (Spillers 65).[189] Similarly, Omi and Winant point to the visual dimension of racialization, i. e., "the extension of racial meaning to a previously racially unclassified relationship, social practice or group," as they term the attribution of racial categories (Omi and Winant 111). The visual is key to these processes. As Omi and Winant argue:

> Bodies are visually read and narrated in ways that draw upon an ensemble of symbolic meanings and associations. Corporeal distinctions are common; they become essentialized. Perceived differences in skin color, physical build, hair texture, the structure of cheek bones, the shape of the nose, or the presence/absence of an epicanthic fold are understood as the manifestations of more profound differences that are situated *within* racially identified persons: differences in such qualities as intelligence, athletic ability, temperament, and sexuality, among other traits. (Omi and Winant 111, emphasis in original)

The mechanisms of racialization ascribe arbitrary significations to features of human bodies. The interaction of the visual and the verbal with the body thus act on the surface of bodies.

Analogous to such racial inscriptions, in Kincaid's works discourses on sex and gender also constitute bodies. While bodies seemingly operate on their own when they change during puberty – producing new parts, for instance, such as pubic hair or female breasts, discourse also attributes meaning to these body parts in a way that means they can never exist in a neutral state. The female body here is always already marked with meaning in the sense of Judith Butler's theorization of how language constructs categories of sex:

> Gender ought not to be conceived merely as the cultural inscription of meaning on a pregiven sex (a juridical conception); gender must also designate the very apparatus of production whereby the sexes themselves are established. As a result, gender is not to culture as sex is to nature; gender is also the discursive/cultural means by which "sexed nature" or "a natural sex" is produced and established as "prediscursive," prior to culture, a politically neutral surface *on which* culture acts. [...] This production of sex *as* the prediscursive ought to be understood as the effect of the apparatus of cultural construction designated by *gender.* (*Gender Trouble* 7, emphasis in original)

189 This is, for instance, also apparent in Frantz Fanon's well-known train episode in which a white boy's gaze ascribes meanings to Fanon's dark skin, which Fanon can neither determine himself nor evade.

In this production of sex through gender, bodies are inscribed by being read as such and cannot be thought as unmarked, as every approach to the body already comes with a culturally predetermined set of readings.

My sketch here of useful theoretical notions of the human body in the context of Jamaica Kincaid's works has so far made quite extensive use of the metaphors of reading and writing in order to understand the body as a medium. In part this follows the terminological choices made by other scholars (Grosz: "inscription," Omi and Winant: "reading bodies," Spillers: "markers," "hieroglyphics"), yet it also reflects my own understanding of how the body functions in Kincaid's works. I contend that the body as a medium in its constitution is largely dependent here on both the visual and the verbal, i. e., on image and text.

However, the following chapters also show that, while the bodies of characters in Kincaid's works undergo and experience such inscriptions, and while the texts accordingly discuss processes that are acted out *on* bodies, the novels' protagonists also use their own bodies to resist such determinations imposed on them by others. Bodies withstand being fixed in categories and, via nonverbal practices, counter meanings ascribed to them. By enacting colonial discourse with differences that speak to the characters' agency, they contest colonial history that is that seemingly is already inscribed on their bodies.

The following analyses will focus on three of Kincaid's novels that center on female bodies as engaging with the world – *Annie John*, *Lucy*, and *The Autobiography of My Mother*. Presenting in-depth readings of each novel in response to specific questions of the female body, each of the sections below, devoted to one of these novels, analyzes passages from the two other to support my interpretations. My aim is to chart how conceptions of the female body develop in Kincaid's texts. While *Annie John* in part exemplifies the devastating effects of an internalized white gaze, *The Autobiography of my Mother* presents more radical performances and rejections of colonial claims on its protagonist's body.

4.1 Transgressions of Gender: The "Unruly" Female Body, Grotesquerie, and Odors

Annie John is the coming-of-age story of its eponymous protagonist. Chronologically narrated in retrospect by an autodiegetic narrator, the novel relates episodes in Annie John's life from ages nine to seventeen. The chapters detail changes in her social life, i. e., in her school education, friendships with her peers, a changing relationship with her parents, and how Annie's physical body goes through the changes of puberty while all of this is happening. Growing a visibly female body is a traumatic experience for Annie. She experiences the development of her own body throughout puberty in terms we might call, with Bakhtin, grotesque. Yet the unruliness of grotesquerie simultaneously withstands social concepts attributed to the female body. This chapter explores the resistances of female bodies in Kincaid's texts to external rules that attempt to bind them together with fixed meanings. Their very physicalities defy discursive constructions of normative gendered behavior, as we see with Annie John's growing body and the way she experiences it as a distorting development. By presenting the changing pubescent body of its protagonist through grotesquerie, *Annie John* highlights the social and discursive inscriptions of gendered bodies in the sense of Judith Butler's approach to gender. "[G]ender is in no way a stable identity or locus of agency from which various acts proceed; rather it is an identity tenuously constituted in time" (Butler, "Performative Acts" 519). Here, such a constitution is eminently marked by colonial ideals.

The first section of this chapter, "Growing the Grotesque Female Body," focuses on the physical changes during puberty that are linked to gendered and racializing projections. Reflecting on her changing self-image, Annie John initially presents an internalized white gaze that seeks to frame her body as strange and undesirable. As she grows up, however, she inverts the grotesquerie of her body into a challenge of the very colonial notions that had hitherto limited her. The pubescent body's transgression of its physical limits thus evolves into transgressions of gendered norms. As I will now show, the binaries of inside and outside, and of norm and deviation, are thus dissolved with and through Annie John's body. The second section of this chapter on "Offensive Odors" will then explore the challenge of racialized and gendered attributions through emissions of bodily odors. Drawing on approaches to race and gender in the new field of smell studies, this section presents the defiance of colonial discourse through the concrete physical body that makes itself known through its pungent smelliness in *Annie John*, *Lucy*, and *The Autobiography of My Mother*.

Growing the Grotesque Female Body

Throughout the novel, Annie John's body is in constant flux. Going through puberty, her body transgresses its own boundaries with the excretion of body fluids and the growing of new appendages. Simultaneously, and because of these physical changes, Annie John learns to relate to the world differently. As her body develops from that of a child to that of a young woman, her immediate society begins to view her differently – people now expect a gendered behavior of her, which irritates her as she learns to navigate these new claims made on her body. The book presents this experience of the changing body and its changing relations to the world as a series of transgressions by the body itself, by the world onto her body and her sense of self, and by Annie John herself, in moments when she crosses the perceived social boundaries imposed on her. This string of transgressions comes to her with a sense of alienation, which is presented by the text through descriptions of her body in grotesque terms. The narrative of Annie's growing up is conspicuously rife with portrayals of her body, its protruding parts and its excretions, such as pubic hair, sweat, and menstrual blood. In Mikhail Bakhtin's terms, "[t]he [grotesque] object transgresses its own confines, ceases to be itself. The limits between the body and the world are erased, leading to the fusion of one with the other and with surrounding objects" (*Rabelais* 303). While Annie John's body indeed changes and thus literally transgresses on the world, I read this as a metaphor for her new status in society as a "young lady" who is expected to behave according to gendered norms. The overt representations of Annie's bodily functions then signify her rejection of such social norms, as I will show in what follows.

Annie's changing body and social position leave her in a state of instability and uncertainty. Her body disrupts normality. To be sure, I do not personally regard any of Annie John's thoughts, feelings, or body functions as "abnormal" or as "grotesque" in a judgmental way (as the term might be understood in everyday language use). Rather, I am interested in Annie's own sense of alienation as presented by the novel, as well as the functions of the grotesque (in the strict sense of the transgressive, exaggerated, and excessive as defined by Mikhail Bakhtin or Mary Russo) as a literary technique to approach and comment on gender categories and the fixity of colonial discourse in general. The grotesque body in *Annie John* questions the fixity of the category of "normality." Mary Russo points out that the feminine and the grotesque overlap in their constructions and positions "as superficial and to the margins" (6). Russo even sees the two terms as doubling: "It might follow that the expression 'female grotesque' threatens to become a tautology, since the female is always defined against the male norm. Indeed, in many instances, these terms seem to collapse into one another in very powerful representations of the female body as grotesque" (12). Moreover, as Justin Edwards and Rune Graulund contextualize, in a "postcolonial context, grotesquerie can highlight 'difference' by identifying old and new spaces of centrality and normalcy, if only to transgress the boundaries that have been established by the forces of a colonial power" (124). In a British (post)colonial context, normality is regulated as "white," "male," and "able-bodied." This would per se define Annie John as deviant, simply by being black and female. Or, as Lizabeth Paravisini-Gebert puts it, "[c]arnival, and what Bakhtin calls the grotesque – become avenues for the articulation of resistance to a dominant culture's tendency to undermine the static, seemingly unshakable hierarchies imposed by colonial rule" ("Carnival" 217).

Jamaica Kincaid's *Annie John* presents the resistance to colonial rule in the learning processes of a girl growing both into her own body and into the body of society, each of them governed by discursive regulations. Annie John's narrative of growing up includes a painful growing apart from her mother. Initially, the text presents the daughter as a double of her parent, a position of which she is soon deprived. The novel begins with the loss of the locus amoenus: at age nine, Annie moves with her family from their home to a temporary dwelling, as their house's roof needs to be replaced. Their home for the summer is located near the town's graveyard, a space that is most alluring to her. The first chapter is thus concerned with death, which anticipates the increasing loss of the idyllic in the following chapters. Simultaneously, the previously unchallenged union of mother and daughter, marked by a sharing of all activities (for instance shopping, cooking, bathing), begins to disintegrate. The novel thus begins with a change from an assumedly innocent childhood, sheltered by the nuclear family, to an awareness of life – and, as the proximity of the graveyard suggests, death – in society beyond the private realm.

This change prominently shows itself in an episode of shopping for a dress, which is necessitated by a growth spurt when Annie John is twelve years old. Annie John perceives the bodily changes that occasion the acquisition of this new garment as an alienation from herself: "My legs had become more spindlelike, the hair on my head even more unruly than usual, small tufts of hair appeared under my arms, and when I perspired the smell was strange, as if I had turned into a strange animal" (*AJ* 25). In this observation of her own body, Annie sees her body as rebellious. It is "unruly"; it acts on its own when new features "appear"; and it is unknown, unfamiliar, "strange" to her. Here, she experiences her body's "unruliness" as directed against herself. This will eventually change, and Annie John will use her rebellious female body as a tool to resist norms imposed on her as a young woman in a colonial society.

What Annie John describes in this early alienating encounter with her own body are the typical changes of human bodies during puberty, both male and female. Physical growth and hormonal changes that effect changes in the chemical composition of body fluids (such as perspiration) are a shared human experience. Annie's observation of her body thus is not out of the ordinary at all. The way in which Annie describes her body, however, casts it in decidedly grotesque terms. In *Rabelais and His World*, Mikhail Bakhtin determines that "[t]he grotesque image reflects a phenomenon in transformation, an as yet unfinished metamorphosis, of death and birth, growth and becoming" (25). The grotesque body then is "a body in the act of becoming. It is never finished, never completed; it is continually built, created, and builds and creates another body" (Bakhtin, *Rabelais* 317). The pubescent body would be a prime example of the grotesque in Bakhtin's conception. It transgresses itself, it reaches beyond its own boundaries by growing parts that extend from it in new ways, such as Annie's spindly legs or her axillary hair. These body parts protrude; they reach towards an outside world, which is for Bakhtin a key feature of the grotesque (see *Rabelais* 316–317). This is later heightened in the female body with the growing of breasts.

In Annie's first description of her pubescent bodily change, the grotesque also manifests itself in her perspiration. Again, a part of the body pulls from the inside to the outside; it transgresses its limits, a movement that is immediately connoted with strangeness. Indeed, Annie's perception of her changing body is so alien to herself that she voices it through another grotesque motif – the metamorphosis from a human into an animal being (see Bakhtin, *Rabelais* 316). Comparing her body with "a strange animal" emphasizes the physicality of herself, which during puberty becomes part of Annie's self-conception.

At the same time, Annie John also experiences the alienation from the body of her mother. She had been her mother's double, following her around and copying her ways of doing things and feeling as an extension of her mother's body. Now, Annie is deprived of this position when the mother refuses to continue performing sameness by wearing the same cloth. "Oh no," the mother says at Annie's suggestion for a role of cloth. "You are getting too old for that. It's time you had your own clothes. You just cannot go around the rest of your life looking like a little me" (*AJ* 26). So far, shared appearance in clothing had signaled familiarity. Now, the change of outward clothing performs an inward change – the emotional separation of Annie from her mother Annie.

Julia Kristeva takes up Bakhtin's notion of the grotesque and adds to it the dimension of gender in her concept of the "abject" as that which falls away from the "I" and, in the process, constitutes the "I" (Kristeva 3).[190] In a thought that resembles Bakhtin's grotesque in that the abject is also that which transgresses the confines of the body, Kristeva locates bodily excretions such as blood, pus, sweat, and feces beyond the border of the "I." "These body fluids, this defilement, this shit are what life withstands, hardly and with difficulty, on the part of death. [...] Such wastes drop so that I might live, until from loss to loss, nothing remains in me and my entire body falls beyond the limit – *cadere*, cadaver," Kristeva writes (3). Moreover, according to Kristeva, an awareness of the processes of abjection finds its origin in the detachment from the mother. As the child enters the symbolic realm in which "I recognize my image as a sign and change in order to signify"

190 For instance, Mary Russo rightly points out that "Bakhtin [...] fails to acknowledge or incorporate the social relations of gender in his semiotic model of the body politic, and thus his notion of the Female Grotesque remains in all directions repressed and undeveloped" (63).

(Kristeva 14), it abjects the mother who has brought it into being, in order to exist outside of her. Before the child becomes a subject, it thrives on mimesis "by means of which [s]he becomes homologous to another in order to become him [or her] self" (Kristeva 13). Through the abjection of the maternal the child enters their own personhood. In Annie John's case, the child feels forced into a symbolic order, in which her own image is laden with significations, when she visibly grows into an adult female body. While she later may actively turn against the union with her mother, the girl initially feels rejected and deprived of her inherited mimetic position close to her parent. Here, the abjection of the maternal is presented as a forced process, initiated by the symbolic inscriptions of the female body which cannot be overcome. Annie's sense of rejection here emphasizes the power of the symbolic, i. e., patriarchal policing discourses, which manifests itself on her body.

The daughter's reaction to the realization that she is now forced to leave the comfort of maternal mimicry is performed by her physical body, which now suddenly is as beyond her control as the choice of cloth in which she might clad her body. The girl abruptly turns hot, then cold, "and all [her] pores must have opened up, for fluids just flowed out of [her]" (*AJ* 26). Elizabeth Grosz points to the menace of body fluids, such as Annie John's perspiration, especially when they refuse to be controlled: "Body fluids attest to the permeability of the body, its necessary dependence on an outside, its liability to collapse into this outside (this is what death implies), to the perilous divisions between the body's inside and its outside. They affront the subject's aspiration toward autonomy and self-identity" (193). Annie John's excessive perspiration could then be read indexically, physically pointing not only to her momentary emotional distress, but also to a disintegration of her perceived selfhood. Her mother's forceful separation through the termination of their visual doubling threatens the girl's very sense of self, as is expressed by the dissipation of her physical barriers to the outside world. Annie John quite literally cannot keep herself together.

Such transgressions of the boundaries of the physical body in a grotesque fashion moreover threaten the notion of order.[191] As Mary Douglas emphasizes, systemic regulations are manifested on the individual body and "[i]t is only by exaggerating the difference between within and without, above and below, male and female, with and against, that a semblance of order is created" (5). In this, the conception of the personal body structurally functions in the same way as notions of "orderly" society, especially in a colonial system which defines its norms against "the Other." Yet, here the seeping of body fluids is less threatening to Annie's sur-

191 Bakhtin explicitly mentions sweating as an action of the grotesque body (*Rabelais* 317).

roundings than to her own position and subjectivity in it. Thus, the grotesque and its emphasis on the physical human body might foreshadow the body's unruly behavior in a societal context, but in this situation early in Annie John's life, it points to the uncertainty Annie experiences in being cast out of the amoenic spaces she inhabited near her mother and as her double.

Annie John's autodiegetic narrative position identifies her mother as the perpetrator of her pain. Her mother, however, cites her daughter's advanced age for her actions. She supposes that her daughter is "getting too old" for their harmonious union and notes that the girl is "on the verge of becoming a young lady, so there were quite a few things [she] would have to do differently" (*AJ* 26). What the mother refers to here would be the onset of puberty, which is visually manifesting on Annie's body. The amoenic is relinquished when her previously infantile body begins to evince the marks of a gendered body in puberty. This physical change is immediately linked here to a gendered social role into which Annie John is forced. Still, while the daughter's perception of her situation positions her mother as an agent of colonizing rules in *Annie John* – very similar indeed to the mother-daughter-dynamic in "Biography of a Dress"[192] – she really only acts as a proxy in a system she cannot escape herself.[193]

Again thinking about the girl's societal position with Kristeva's concept of maternal abjection, it becomes clear that the detachment process of this familiar relationship, which *Annie John* presents as a visual upending of sameness and which signifies the transcendence of the semiotic order, indeed points to the girl's entering of the symbolic order, which in Kristeva's term is tied to the realm of the father (13).[194] The figure of the mother now emerges as split, which is mirrored in the differing interpretations of the mother by various scholars. Some scholars read

192 See chapter 2.3, section "A Yellow Dress and a Brown Girl."

193 This system is also marked by patriarchy. While patriarchy perhaps does not immediately emerge as a topic at hand as the majority of conflicts in *Annie John* are fought between mother and daughter and the father figure plays only a secondary role, the "rule of the father" nonetheless provides the basis for such conflicts throughout. Annie John's mother thus acts as an enforcer of both colonial and patriarchal rule, which are two sides of the same coin. See Moira Ferguson, *Jamaica Kincaid: Where the Land Meets the Body*, esp. chapter 2 "Annie John: Mother-Daughter – Christopher Columbus," for an analysis of patriarchy in *Annie John*.

194 With the terms "semiotic" and "symbolic," Kristeva distinguishes two aspects of language and development of subjectivity. Kristeva denotes the symbiotic state of mimesis of child and mother as "semiotic," which is determined by emotion. In relation to the variability of emotion, semiotic language features a rejection of denotative fixity. After the Lacanian mirror-stage and after maternal abjection, the "symbolic," in Kristeva's terms, represents the patrilineal. Symbolic language then follows the principle of the patriarchal law, which relies on denotation. For a detailed discussion of this terminology, see also Judith Butler's essay "The Body Politics of Julia Kristeva."

the mother-daughter relationship in *Annie John* as empowering to the girl,[195] while others read the mother as an extension of the oppressing colonizer.[196] The reason for this division in scholarly approaches to the mother figure in *Annie John* probably lies in the split within the character's significations itself. Chinmoy Banerjee also assesses the changing mother-daughter relationship as a change in the conception of the mother figure: "With Annie's splitting of her mother, the two mothers become politically antithetical, the cast-out bad object being identified with the colonial patriarchal power and norms and the introjected good object being identified with opposition and resistance" (40–41). While in my interpretation here the mother represents an alignment with the colonizer's ideals to the daughter, I generally do not intend to restrict the reading of the mother to this position. Rather, while keeping in mind the doubling of the mother figure (as both representation of colonizer and as empowering ally), I here foreground the mother's role as restricting and imposing colonizing significations on her daughter solely in the context of Annie John's changing body and society's new impositions on it.

Naked, Annie John beholds herself in a mirror:

> I was so long and bony that I more than filled up the mirror, and my small ribs pressed out against my skin. I tried to push my unruly hair down against my head so that it would lie flat, but as soon as I let it go it bounced up again. I could see all the tufts of hair under my arms. And then I got a good look at my nose. It had suddenly spread across my face, almost blotting out my cheeks, taking up my whole face, so that if I didn't know it was me standing there I would have wondered about the strange girl – and to think that only so recently my nose had been a small thing, the size of a rosebud. (*AJ* 27)

Individual body parts and the body as a whole are exaggerated in their shapes and sizes in Annie's self-perception. She separately pays attention to individual parts of the body, verbally detaching them from the whole. They are described as disproportionately large, extending and protruding beyond Annie's body. The nose, which Mikhail Bakhtin likewise describes as one of the key grotesque features of the face (*Rabelais* 316), attains excessive size and takes over her visage. What is at play in this exaggeration within the mirror scene is Annie's double vision of herself. The focus on bouncy hair and a broad nose recalls racist stereotypes of Black people, which exaggerate just these features (among others, such as lips or eyes). Disproportion is a key component of racist caricature of Black bodies, as perpetuated for instance by cartoons and collectibles.[197] In centering on hair

195 For instance, Donna Perry ("Initiation" 134) or Verna Ena George (40–43).

196 For instance, Colena Gardner-Corbett (75), Laura Niesen de Abruna (281), or Simone Alexander ("I am Me" 52).

197 See for instance Patricia Turner, *Ceramic Uncles and Celluloid Mammies.*

and nose and in describing them in exaggerated terms, Annie applies a racist white gaze to her own mirror image, doubling it in the sense of Du Bois and Fanon,[198] which casts her image in that of a "strange girl."

More problematic even than Annie John's growth is her body's ability to change on its own. Annie John is not in control here; her body has taken over. Her unruly hair cannot be flattened, and her later musing that she might screw herself into a set of clamps at night in order to combat her physical growing (*AJ* 27) might be read as a sign of the girl's desperation to come back into agency and to govern her own body. At the same time, it signifies her desire to conform to parameters of colonial standards of beauty. She has as little power over her physical body as over norms of beauty, and with that, over the colonial ideals that reveal themselves in her own mirrored gaze.

Annie's experience of bodily alienation culminates in the onset of her menstruation, as the presence of bodily fluids becomes a horror scenario. She experiences this as yet another, new strangeness that manifests itself in and on her body, and as one which moreover is laden with shame (*AJ* 51–52). Not only does her body transgress its borders in new ways, but she suspects that it even transmits this fact to the outside world: "I was sure that everything about me broadcast, 'She's menstruating today. She's menstruating today'" (*AJ* 52). This assumption that the body is operating on its own accord again highlights Annie John's loss of control. Grosz connects female monthly bleeding with an ontological instability of womanhood:

> [M]enstruation, associated as it is with blood, with injury and the wound, with a mess that does not dry invisibly, that leaks, uncontrollable, [...] indicates the beginning of an out-of-control status that she [woman] was led to believe ends with childhood [and its involuntary soiling of oneself and one's clothes]. [...] This necessarily marks womanhood [...] as a paradoxical entity, on the very border between infancy and adulthood, nature and culture, subject and object, rational being and irrational animal. (205)

Grosz's observations here emphasize the enduring state of transgression in which women are cast based on their bodily functions, which frames women in a position of ongoing grotesquerie.

In Annie John's case, the image of the bleeding wound indeed is so threatening to the girl that it causes her to faint. The blackout once again demonstrates Annie's current powerlessness and inability to control her own body. Her syncope occurs at school and the nurse quickly chalks it up to "the fright of all the unexpected

---

198 See the section on "A Yellow Dress and a Brown Girl" in chapter 2.3 for details on the internalized white gaze as theorized by Du Bois and Fanon, as well as a detailed analysis of double-consciousness in Kincaid's "Biography of a Dress."

pain" (*AJ* 52), but the girl realizes that "I had fainted after I brought to my mind a clear picture of myself sitting at my desk in my own blood" (*AJ* 52). The imagined scenario links two horrors: the shame of public display of the abject, i. e., of menstrual blood flowing from her body, and the equally horrifying thought of bleeding out and dying without a direct cause. The body's grotesquerie here is characterized by its perceived acting on its own, which renders the girl utterly without agency.

This changes, however, when in Annie's group of peers, menstruation becomes a thing of value. While her changing pubescent body means the forced entry into a world ruled by gendered norms, Annie John also gains a new community – that of her female schoolmates. She enters a homosocial group of peers with whom she shares experiences of their changing bodies, such as the growing of breasts, which they perceive as an increasingly desirable development of their young bodies (*AJ* 74, 80). The novel's protagonist, who enjoys being at the center of her friends' attention, gains recognition by being the first to menstruate. Her best friend Gwen is moved to tears by Annie's revelation, but she is not crying out of sympathy for Annie's bodily horror, but for herself, as she had not yet experienced her menarche (*AJ* 52). Annie achieves further admiration when she presents her bloody sanitary cloth to her friends and shares her newly obtained knowledge of female monthly bleeding. For her, then, the admiration of her peers and her new knowledge render the previously daunting grotesquerie of her own body into a tool of power.[199]

Kincaid's later novel *Lucy*, also evokes menstruation in a negotiation of power when the protagonist bleeds upon having penetrative sex for the first time during puberty. Realizing the presence of blood, her partner smiles and says "'Oh,' a note too triumphant in his voice" (*Lucy* 82). Refusing to be perceived as a special conquest, Lucy quickly notes that this blood would simply be her period coming on – no need for "triumph." "I did not care about being a virgin and had long been looking forward to the day when I could rid myself of that status, but when I saw how much it mattered to him to be the first boy I had been with, I could not give him such a hold over me" (*Lucy* 82–83). Lucy here negates the hold of body discourses on her in a two-fold way. First, she rejects the notion that female virginity might be desirable; sexual intercourse here is a way to defy this claim on her own body by preempting it. And second, she interprets the presence of blood

---

199 One might similarly interpret the empowering experience of purposefully producing fart noises at a lesson in etiquette to which Annie John's mother sent her. The girl here disregards the instruction in colonial conventions and opts to make her peers laugh instead (*AJ* 27–28). Imitating sounds of the bodily grotesque, Annie resists the colonial power's attempted influence on her body, here literally, as the emulation of fart sounds takes place during curtsy lessons. See also Valens 126.

as something mundane, less exciting than proof of her "deflowering." By citing menstruation, Lucy thus refuses the discursively fabricated exceptionality of this sexual act. In *The Autobiography of My Mother* menstrual blood even has magical properties, enabling women to charm men, as Lise LaBatte does with Jacques by mixing her blood into his food. He promptly becomes her husband (*ABM* 65). Through Kincaid's oeuvre, menstruation thus evolves from a fearful experience to female potency.

In *Annie John*, the powerful aspect of the grotesque body with its unruliness and transgression as significations of agency is also indicated by Annie's own desires in her friendships with other girls. Throughout the book, Annie entertains three close relationships with girls of the same age – Sonia, Gwen, and "the Red Girl." For these girls Annie professes deeply felt love, which expresses feelings read by some scholars as beyond girlish friendships and rather as a queer desire for girls.[200] Considering the colonial patriarchal society this takes place in, Annie's taking a homosexual love interest in her female friends is the performance of a social transgression in itself. Keja Valens articulates this very clearly when she writes that

> [t]he pursuit of desire between girls in *Annie John* is intertwined with anticolonial struggle, for it undermines colonial heteronormativity. The roots of heterosexuality in the Caribbean certainly reach beyond the history of conquest and colonialism, but while male-female relationships are extremely common throughout the Caribbean, they do not always take on the same institutional or symbolic roles that they do in Europe, more specifically in colonial England. The insistence on heterosexuality as the norm that can and must not be violated – and its concomitant regulation of the boundaries of gender and family roles – belongs to a Victorian morality whose imposition forms part of British colonialism. (124)

In this, Annie's desire for female romantic relationships signals the girl's departure from both heterosexual and homosocial norms.

Such a bypassing of social norms is also indicated by the specific girls in which Annie John is interested. While the narrator (and internal focalizer) Annie John describes all three as "beautiful," the details of the physiques in their differing from "normality" rather mark the girls as social outsiders (Sonia and the "Red Girl" much more so than Gwen). Annie's attraction to difference is evident early on, with her earliest school friend, Sonia, whom Annie – without qualms – de-

200 See for instance Valens or Strongman. A number of other scholars choose to read Annie's fixation on her girlfriends as replacements for the previous loss of the close relationship with her mother (see, for instance, James Nagel or Diane Cousineau). While this interpretation suggests itself by the temporal order of the loss and gain of female significant others in *Annie John*, this does not exclude a queer desire from Annie's future relationships.

scribes as a dunce. Her other schoolmates frown at Annie's interest in Sonia. Nonetheless Annie notes, "I found her beautiful and I would say so. She had long, thick black hair that lay down flat on her arms and legs" (*AJ* 7). Sonia's physical deviation through excessive body hair delights Annie; she enjoys interacting with the mass of hair on Sonia's body (*AJ* 7).

This delight in the differing body is repeated when Annie attends a new school. Here she meets Gwen and immediately falls in love with her. The scenes with Gwen are described as moments steeped in beauty, rays of morning sun illuminate the first meeting of the girls. Annie sees Gwen in terms of perfection. Her school uniform is orderly; her shoes are polished. Yet Gwen's hat, adorned with a satin ribbon, sits on her head lopsidedly (her head is too small for it); the hair ribbons hold together "shrubby" hair; her "bony knees [...] were always ash-colored, as if she had just finished giving them a good scratch"; and her lips were "the shape of a saucer broken evenly in two" (*AJ* 47). Annie's description of her girlfriend make clear that what signals perfection to her is phrased in terms of disruption – the crooked, the unruly, the scratched, the broken. With her appreciation of difference, Annie John's ideals of beauty have significantly shifted from the internalized white gaze she had previously applied to her own mirror image. Bones sticking out, bushy hair, and wide round lips now are not condemned, but appreciated.

Valens asserts that "Gwen's features are also distinctly Caribbean (shrubby hair, ash-colored knees, flat nose, wide lips, high cheekbones) [which] suggests that Kincaid elaborates a Caribbean erotics of the grotesque, which refuses colonial norms of beauty and instead embraces the negative stereotypes of Afro-Caribbeans as precisely their most beautiful and desirable qualities" (131). Reading Annie's new ideals of female beauty as transgressive of colonial standardized notions of beauty transforms the female grotesque body into the desirable body. The shift in desirability thus decenters the colonizer's power to police discourses of beauty and restores individual agency to Annie John.

The appreciation of nonconforming beauty becomes even more pronounced when Annie John makes the acquaintance of the "Red Girl," as Annie calls her. They meet when the Red Girl offers to pick guavas Annie cannot reach and the Red Girl expertly climbs the guava tree in the fashion of boys (*AJ* 56). Having been taught not to climb but to solely throw stones in the hope of hitting loose the desired fruit, Annie instantly admires the Red Girl, who makes use of a much more efficient method to achieve her goal. The interaction between the two girls echoes a conservative scene of heterosexual courting: The Red Girl here performs the role usually assigned to a boy, freely transgressing the boundaries of gender roles, whose pressures Annie feels all too keenly. Metaphorically, this episode might then be read as a comment on gender norms in general: the

Red Girl refuses to adhere to a gendered male advantage in a patriarchal society, which holds back girls and women with rules of propriety. It is this disregard of gendered expectations that attracts Annie John to her new friend.[201]

In the course of *Annie John*'s 148 pages, the conception of the transgressive female body changes from threatening the one who inhabits it to threatening gendered societal norms. Initially the gendered female body is a problematic discovery for Annie John, but while she grows into it, she learns to defy the restrictions placed on her body by colonial society. In this, Annie John begins to regard her body as a tool to assert herself in the world. This is most obvious at the very end of the book, when Annie John's separation from her mother is completed by a spatial separation. Growing up culminates in Annie's leaving the island by ship in order to start nursing school in the colonizing mother country Great Britain. The now seventeen-year-old girl is devastated at the prospect of leaving that was pushed by her mother, and Annie experiences this pressure as yet another betrayal. Simultaneously, she delights in redirecting the pain of separation to her mother, as she secretly vows never to return to her home in the West Indies (*AJ* 132–133). With this, the text has come to a completion of its story of separation of mother and daughter. Yet, this seeming linearity is disrupted by the circularity of return to the texts, beginning when Annie John again becomes the physical double of her mother: standing by the ship's rail, Annie John looks back at her mother and her former life, thinking of her mother's hypocrisy of professing her love for her daughter while arranging for their separation. Annie regards her secret plan not to return to her mother as retribution, but in this she repeats her mother's perceived patterns of action: "So now I, too, have hypocrisy, and breasts (small ones), and hair growing in the appropriate places, and sharp eyes and I have made a vow never to be fooled again" (*AJ* 133). Her body now has changed from that of a girl to that of a woman as symbolized by the breasts protruding from her chest. The female body now indicates equality to a previously oppressing force as symbolized by her mother. What had appeared horrifically grotesque to her before, now indicates the weaponry she now possesses to challenge others, and which she may use to find power.

In *Annie John*, the textual presentation of the grotesque body both highlights and questions binary structures. Grotesque physical transgressions point to the existence of borders between inside and outside, body and world, the normal and the nonconforming, etc. By presenting such binary structures through transgressing them, the book both reveals and challenges them. Ambivalence, according to Bakh-

---

201 This is furthermore supported by Annie John's admiration of the Red Girl's state of unkemptness. See the following section on how odors and dirt upset colonial order in Kincaid's texts.

tin, is one of the grotesque's indispensable traits. "For in this image we find both poles of transformation, the old and the new, the dying and procreating, the beginning and the end of the metamorphosis" (Bakhtin, *Rabelais* 24). This simultaneity through transgression and transformation decenters the very binarity of the two poles. As Justin Edwards and Rune Graulund put it, "the grotesque offers a creative force for conceptualizing the indeterminate that is produced by distortion, and reflecting on the significance of the uncertainty that is thereby produced. This means that the discombobulating juxtapositions and bizarre combinations found in the grotesque figures in literature and the other arts open up an indeterminate space of conflicting possibilities, images and figures" (3). The presentations of female bodies in grotesque terms in *Annie John*, I conclude, thus refuse the totality and fixity of colonial logic and open up conceptions of gender and normativity to a multitude of realizations.

Upsetting Colonial Order: Offensive Odors

Besides growing grotesque parts, bodies in Kincaid's texts also transgress their boundaries by emitting odors. While my analyses largely focused on visual aspects of the female body in the previous section on grotesquerie, also other senses take part in the negotiations of gendered and racial attributions in Kincaid's works. To emphasize the importance of senses, Mark Smith in his monograph *How Race Is Made* notes that "[e]ven though we know that 'race' is a construct, an invention and category that defies scientific verification, we still understand that construction as a largely visual enterprise. [...] But the preference for seeing race is as much a social construction as 'race' itself" (2).[202] Generally, other senses, such as smell, hearing, and touch participate in the making of race, although their mechanisms might be less conspicuous. Isabel Hoving rightly argues that by including the sense of smell with body conceptions, Kincaid's writing takes up a form of knowledge production that turns away from theorizing the body in terms of text or through textual metaphors (such as suggesting bodily fluids as evocations of ink), which in itself challenges Eurocentric forms of knowledge production (Hoving 226–227). In Kincaid's works, bodily smells thus become ways to engage with and challenge the discursive notions that take up odors in othering mechanisms, as the following section explores.

202 Similarly, Andrew Kettler holds that "[r]acism is one of the many aspects of culture and language that is felt through the sensory organs" (213).

The protagonist of *Lucy* very well understands bodies as communicative media which may transgress what is obviously visible, as is evident in her perception and evaluation of body odors. For Lucy, odor becomes a means to assert oneself in the world. Thinking about Mariah, Lucy's employer, close friend, and substitute mother figure, Lucy regards her as a paragon of privileged white upper-class American women who are unaware of the ease of their own lives and who hence never had to develop any confidence to face the difficulties of life (*Lucy* 26–27). Considering Mariah's seemingly effortless physical appearance and beauty, Lucy notes that "[t]he smell of Mariah was pleasant. Just that – pleasant. And I thought, But that's the trouble with Mariah – she smells pleasant. But then I already knew that I wanted to have a powerful odor and would not care if it gave offense" (*Lucy* 27). Lucy evaluates Mariah's appearance and presence in general as "pleasant," which is mirrored by her perceived odor. "Pleasant" here, however, is not connoted positively, but rather to be read as "vapid" and "inexpressive." It does not contribute to an active presence. Lucy, in contrast, wants her own odor to announce her to her surroundings and to make her presence known. Scholars have variously read Lucy's desire for a strong odor as empowering.[203] Here, I read Lucy's stance on odorous bodies in the context of two other novels by Kincaid – *Annie John* and *The Autobiography of My Mother* – and thus delineate a development in the conception of body smells in Kincaid's works and integrate this development into a reading of the physical body as engaging with colonial history through its odorous transgressions of order.

Lucy desires for her body to extend beyond its confines and to assert itself in the "outside" world via its odor. In Kincaid's previous novel, *Annie John*, the much younger protagonist locates her own body odor in the realm of the grotesque; as I expounded, she notices it simultaneously with her body growing beyond itself during puberty, which she experiences as frightening. Annie John finds her odors "strange" (25) and alienating. Lucy understands her bodily smells as part of herself that grows out from her body. In this transgressive quality of odors, Lucy repeats Annie John's notion of the grotesque. Yet the protagonist of the later novel does not regard this as an estrangement from herself; rather she sees its potential to "give offense" to others.

---

203 See for instance Marquis 100, Hoving 220, or Holcomb 308. In contrast, J. Brooks Bouson surprisingly reads Lucy's stance towards odor in a judgmental context of shame. Bouson equals Lucy's wish to be "dissmelling" with being a "dirty individual" (Bouson 74) and argues that this points to her regarding herself with contempt. I disagree with Bouson and rather understand Lucy as making use of her own body to defy notions of shame that are imposed on her and which she does not support herself at all.

Although Annie John was offended by her own smell, the earlier novel anticipates the pleasure this smell may also provide, like it does to Lucy, as Annie begins to appreciate difference. For instance, the Red Girl's transgression of gender boundaries that Annie John so admires is mirrored by her physical appearance, which equally challenges societal rules.

> The Red Girl and I stood under the guava tree looking each other up and down. What a beautiful thing I saw standing there before me. Her face was big and round and red, like a moon – a red moon. She had big, broad, flat feet, and they were naked to the bare ground; her dress was dirty, the skirt and blouse tearing away from each other at one side; the red hair that I had first seen standing up on her head was matted and tangled; her hands were big and fat, and her fingernails held at least ten anthills of dirt under them. And on top of that, she had such an unbelievable, wonderful smell, as if she had never taken a bath in her whole life. (*AJ* 57–58)

The girl is revered by Annie John for her seeming carelessness at her own deviant appearance. She refuses to adhere to norms of cleanliness with her dirty and torn dress, her unkempt hair, and the smell of her unwashed body.

Annie learns that the Red Girl's mother does not insist on daily bathing, as her own mother does, and she determines, "Oh, what an angel she was, and what a heaven she lived in!" (*AJ* 58). From an adult perspective, Annie John's descriptions of her friend might speak less to the girl's desirable independence, and more to her family's poverty and a possible case of child neglect. Nonetheless, from Annie's youthful perspective, the Red Girl's bodily deviance is enticing in expressing unruliness. Annie John's recognition of the state of the Red Girl's body and dress as "dirty" shows her awareness of societal hygienic standards, as well as her realization that the Red Girl is transgressing them.

Mary Douglas's inquiry into concepts of pollution and belief systems in *Purity and Danger* argues that "[t]here is no such thing as absolute dirt: it exists in the eye of the beholder. If we shun dirt, it is not because [...] our ideas about disease account for the range of our behaviour in cleaning and avoiding dirt. Dirt offends against order" (2). Understanding dirt as yet another discursively defined entity, it becomes clear that its perceived negative aspects indicate not the threat of disease or decay – be it physical or, in a metaphorical sense, moral – but that what is regarded as dirt simply signifies a transgression of what is defined as "normal," i.e., of order. "In chasing dirt, in papering, decorating, tidying, we are [...] re-ordering our environment, making it conform to an idea," Douglas clarifies (3). It is this bodily defiance of conformity that attracts Annie John to her new friend.

The appreciation of body odor is taken even further in *The Autobiography of My Mother*. Initially, human odor is also introduced to the protagonist as something that offends the sensibilities of others:

> My father's wife showed me how to wash myself. It was not done with kindness. My human form and odor were an opportunity to heap scorn on me. I responded in a fashion by now characteristic of me: whatever I was told to hate I loved and loved the most. I loved the smell of the thin dirt behind my ears, the smell of my unwashed mouth, the smell that came from between my legs, the smell in the pit of my arm, the smell of my unwashed feet. Whatever about me caused offense, whatever was native to me, whatever I could not help and was not a moral failing – those things about me I loved with the fervor of the devoted. (*ABM* 32–33)

The young protagonist Xuela, here seven years of age, connects her smell with her humanity. Soap, the common tool of odor erasure, is associated with the colonizing mother figure in *Annie John* and with the epitome of white womanhood, Mariah, in *Lucy.*[204] In *The Autobiography of My Mother* the act of washing oneself is tied to the erasure of one's selfhood. The linking of odor and identity – be it individual, gendered, or racialized – takes up discourses of smell and exclusion, as the example from *The Autobiography of My Mother* suggests. Both *Lucy* and *The Autobiography of My Mother* point to gendered and racialized notions of smell when the texts' Black female protagonists desire odorous bodies, either in opposition to white womanhood or to preserve their identity.

Odors have a subversive effect on Eurocentric knowledge structures precisely because of their marginal position in a hierarchy of senses in the Euro-American culture of knowledge. Since the era of Enlightenment, smell is repressed by sight as the sense of knowledge acquisition. Sight moreover "became associated with men, who – as explorers, scientists, politicians or industrialist were perceived as discovering and dominating the world through their keen gaze" (Classen et al. 84). If sight is associated with power and masculinity, smell becomes an exclusionary factor to relegate "the Other" from the center of power to its margins. Classen, Howes and Synnott point out that Western olfactory discourse often assumes that the "Other" exudes a smell, while the self – which here is defined as white and male – represents olfactory neutrality (161, see also Hyde 56). Andrew Kettler notes that ascriptions of pungent odors to African bodies emerged during the seventeenth century and from then on informed English sensory culture (xi). Thomas Jefferson's remarks on Black people in his *Notes on the State of Virginia* (1787) exemplify such racist attributions of body smells: "Besides those of colour, figure, and hair, there are other physical distinctions proving a difference of race. They have less hair on the face and body. They secrete less by the kidneys and more by the glands of the skin, which gives them a very strong and disagreeable odour. This

---

204 See also my reading of "Biography of a Dress" in chapter 2.3, "Writing between and across Genres," which contextualizes the role of soap within colonial regulations of Black bodies.

greater degree of transpiration renders them more tolerable of heat, and less so of cold than the whites" (139). Similarly, sociologist Georg Simmel writes over 120 years later that "[t]he exclusion of the Negro from high society in North America appears to be due to his body odor, and the complex and deep mutual aversion [of] Jews and Germans has been attributed to the same cause" (qtd. in and trans. by Hyde 56).[205] In the typical fashion of racist ascription, Simmel here assumes that the attributed body odor of African American people, thought to be unbearable to white Americans (Simmel even regards this to be impossible to overcome – his word is "Unüberwindlichkeit," 657), is an inherent quality of a Black body.[206]

It becomes clear that, as Andrew Kettler puts it,

> the shifting of odor upon the racialized other is a timeworn tradition that was expanded within the Atlantic World to provide narratives of disease, pollution, miasma, and labor to justify numerous reprehensible colonialist and capitalist trajectories that situated and disciplined African bodies into a governable discursive and biopolitical space for commodification and political dominance during the Early Modern Era and well into the Scramble for Africa. (xx)

"Odors" in this context are often not even perceived olfactorily, but "smell" functions as a category to exclude, frequently on the basis of an initial *visual* racialized perception (see Classen 80). Odor in racist contexts might hence be regarded as a discursively constructed category, just as any visual category of racialization (as for instance noted by Omi and Winant 111), with which the sense of smell might intersect. "Mixing the sight of a black body with a vast cultural motif buried deep within the sensory membranes, groups of racially incentivized Englishpersons, Western Europeans, and Americans smelled Africans as pungent regardless of the material of bodily odors," Kettler contends (219), emphasizing the intrinsic intersection of discourse and the senses. Bias can delude the nose.

Structurally, such "discursive racist smelling" mirrors olfactory conceptions of the female body. As "others" to the white male, women are also attributed with odors, either as foul-smelling sluts or as sweetly perfumed maidens (see Classen

---

205 German original: "Die Rezeption der Neger in die höhere Gesellschaft Nordamerikas scheint schon wegen der Körperatmosphäre [also termed "Geruchssphäre" by Simmel a few lines above] des Negers ausgeschlossen, und die vielfachte dunkle Aversion von Juden und Germanen gegen einander hat man auf dieselbe Ursache geschoben" (Simmel 657).

206 For further discussion of racist odorous definitions of Black bodies, see Mark Smith 12–16 or Andrew Kettler. Moreover, this asserted body characteristic is neither only applicable to one ethnic group nor a thing of the past. Attributions of supposed odor are still used to exclude marginalized groups, such as immigrants today. See for instance Largey and Watson (31) on the odorous stereotypes white British people voice towards Pakistani Londoners.

87–90). The assumed malodor of women who are regarded as sexually deviant by societal policing of the female body, for instance shows itself in the etymology of the Spanish term for prostitute – *puta* – as Classen points out: "[P]uta is based on the Latin word for putrid, as are similar terms in other modern Latin languages. This stench is the metaphorical product of prostitutes' and sluts' failure to regulate their bodies in accordance with cultural norms and it signals their symbolically polluted and polluting status" (87). The word's history plainly shows the correlation of the evaluation of bodies regarded as outside of the standards of normality (here via sexual practices) and the attribution of certain body odors.

Via their attributed odors, in this olfactory structure, women and people of color are moreover equated with animality in colonial and patriarchal discourses. Classen, Howes, and Synnott find this intersection of odor and notions of the animal in nineteenth-century evolutionary theory, such as that of Charles Darwin, which "postulated that humans lost their acuity of smell in the process of evolving from animals. The marginalization of smell in human society, therefore, appeared necessary of evolutionary and cultural progress, while any attempt to cultivate smell would signify a regression to an earlier, more primitive state" (89–90). As the signifier of "primitivity," smell became associated with a conglomerate of "Others," including people of color and women, as well as people of lower classes – all of whom were linked to a more animal-like state in this logic of attributing and evaluating body odors.

Odorous body characterization provides a reflection of imperialist power structures. While those at the center of power – white men – would be unmarked and represent olfactory neutrality, all those inhabiting the periphery of this power system would be assumed to exude body smells – be they malodorous or pleasant scents. The sheer possibility of being smellable here excludes from the position of power (see also Classen et al. 161).

It is not only in sociology that attributions of odor function as an exclusionary device; literature, too, has long picked up on the motif.[207] And considering that body odor and its perception are located between the physical body and the verbal, since – as shown – odors are as discursively constructed and assumed as any other racial, gendered, or classed category, it is unsurprising that smell finds its space within textual media, as counterintuitive as this might appear.

---

207 See esp. Daniela Babilon's study *The Power of Smell in American Literature: Odor, Affect, and Social Inequality*, or, for instance, references to odor in literature in Classen 81–82 and Largey and Watson 31. See also Larry Shiner's monograph *Art Scents: Exploring the Aesthetics of Smell and the Olfactory Arts*, which goes beyond text and considers odors also in theater, film, music, and performance art, as well as perfume as an art itself.

As the above excursion into the field of smell studies has shown, materialized odors may be superseded by their significations, which are inseparable from visual signs and always already discursively charged with meaning. In this field of claims and ascriptions, actual odors and physical reality are as much entangled with discourse as they are virtually overlayed by the verbal. With Mark Graham, I understand smell – both in literature and without – not as essential and unequivocally ascertainable qualities of bodies (as for instance in Simmel's notion), but rather as performative. Reading smells in a queer context, Graham understands odors in structural parallel to Butler's theorizations of gender. As gender and desire "are the results of social practices and not eternal essences, [they] are open to contestation, they are inherently unstable and they are always in a state of actual or potential flux and transition," so are smells, too, as Graham points out (308).

Understanding odors as performative then brings the physical and the discursive together, which bears the potential to subvert the verbal ascriptions to smells. In the case of Kincaid's protagonists, who so love their own odors, the texts do not question the smelliness of their bodies. In all three cases discussed here – *Annie John*, *Lucy*, and Xuela in *The Autobiography of My Mother* – the bodily odors are detected as powerful by those who emit them. While the younger Annie John finds this disconcerting, the two later protagonists appreciate their own smells for this quality. Comparing the protagonists' attitudes towards odor in Kincaid's three novels, it becomes clear that they develop significantly. This becomes especially clear when the mirror scene, which elaborated young Annie John's growing uneasiness with her changing body and her framing herself with a white gaze in the earlier novel, is repeated in *The Autobiography of Mother* with a difference. Here, the female protagonist reflects on her transitioning body while watching her own image in a mirror:

> I used to stare at myself in an old piece of a broken looking glass I had found in some rubbish under my father's house. The sight of my changing self did not frighten me, I only wondered how I would look eventually; I never doubted that I would like completely whatever stared back at me. And so, too, the smell of my underarms and between my legs changed, and this change pleased me. In those places the smell became pungent, sharp, as if something was in the process of fermenting, slowly; in private, then as now, my hands almost never left those places, and when I was in public, these same hands were always not far from my nose, I so enjoyed the way I smelled, then and now. (*ABM* 58–59)

While Annie John had mused about forcefully keeping her body from changing (with a contraption in which she might screw herself), Xuela never doubts her own image or odor. Taking into account Annie John's double vision of herself, which is informed by the internalized colonizer's gaze, Xuela's contrasting self-love speaks to a much weaker influence of colonial ideals of the body and of beauty on the girl's self-image. They might only play a role in that she defies them by *not* doubting herself and by accepting herself as she is. Annie John's association of body smell with "a strange animal" (*AJ* 27), recalls racialized and gendered notions of the body smells as abnormal, as elucidated above. Hence for Annie John, her bodily odor adds to her body's alienating grotesqueness, which, as her double-consciousness points to her yet uncritical repetition of colonial ascriptions. In contrast, Xuela explicitly connects her own odors with her "nativity" and her "human form" (*ABM* 32–33).

This leads me to conclude that the smelliness of the bodies in these novels resists the colonial and racist discourses that attempt to assign Black female bodies with negative odorousness.[208] The body smells here defy discursive ascription through their physical presence. Pungent and offensive odors in Kincaid's works seemingly negotiate the stereotype of the foul-smelling Other especially since Mariah as the paragon of white womanhood smells pleasantly. Yet Lucy and Xuela turn the stereotype on its head when they adore their own odorous offensiveness. In this, I maintain, their odors perform the pungent Other with a difference that refutes racist ascriptions to their bodies. Bodily odors here signal humanity instead of animality, as they would in racist and patriarchal notions of the smelly body. Facing a discursive ascription that seeks to exclude them from humanity, Lucy and Xuela celebrate the reality of their human forms – their odors and their love of their own bodies thus challenging colonial categories of order.

4.2 Scripted Sexualities: Revisions of the Female Body Through Sexual Practices

Bodily doing in Kincaid's works is often centered around sexuality. The texts' female characters frequently engage in erotic acts. They do so from a young age on, such as the kissing and pinching in *Annie John*. They are also frequently confronted with discourses of sexual morality that attempt to govern their sexual and reproductive bodies. As this chapter shows, scenes in Kincaid's works of sexuality and female procreation (or rather its forced termination) highlight the historical dimension of violent regulation and exploitation of colonized bodies. At the same time, the female protagonists' bodies subvert such claims through symbolic performances of emblematic scenes of colonial sexual violence.

---

208 Kettler (155–194) shows how historically Black enslaved people have long used odors and scents in strategies of resistance to slavery, such as masking one's own odor to throw off blood hounds (see also M. Smith 35). In contrast, Kincaid's contemporary texts celebrate the characters' *own* natural smells in contexts of resistance.

The regulation of sexuality in Kincaid's works is easily read as another yoke of the colonizer.[209] Building on Michel Foucault's theory of the discursive deployment of sexuality,[210] Ann Laura Stoler spells this out when she contextualizes bourgeois and racialized sexuality. Stoler points out that the construction and maintenance of "'Europeanness' was not a fixed attribute, but one altered by environment, class contingent, and not secured by birth" (*Race* 104). It hinged on a successful performance including race, class, and a "normalized sexuality" (Stoler, *Race* 105). Racism and gender are fundamentally intertwined, Stoler argues, as "[t]he very categories 'colonizer' and 'colonized' were secured through forms of sexual control" (*Carnal Knowledge* 42). As such, women's bodies became a central site on which colonial ideals were projected, manifested, or denied. Anne McClintock describes this in terms of figuration when she states that "[c]ontrolling women's sexuality, exalting maternity and breeding a virile race of empire-builders were widely perceived as the paramount means for controlling the health and wealth of the male imperial body politic, so that, by the turn of the century, sexual purity emerged as a controlling metaphor for racial, economic and political power" (*Imperial Leather* 47). Taking this up, I read the textual representations of sexuality in Kincaid's texts as signifying power relations that are integrated with a wider framework of the texts' engagements with colonial history. "[S]exual control was more than a convenient metaphor for colonial domination. It was a fundamental class and racial marker implicated in a wider set of relations of power," as Stoler puts it (*Carnal Knowledge* 45).

Bodies in Kincaid's texts continually challenge sexual control that others seek to establish discursively. Bodily doing in the form of sex extends beyond the immediate action and always simultaneously engages with power structures. The younger protagonists in "Girl," *Annie John*, and *Lucy* negotiate admonitions of "sluttishness" as voiced by their mothers, who here represent colonial ideals.[211] For instance, Lucy's enjoyment of her own sexuality and her taking pleasure in changing sexual partners subsequently contests such attempts at regulating her. Her quest for independence from her overbearing mother is expressed through her rejection of sexual control, articulated in letter to her mother: "I reminded

---

209 See for instance also Gardner-Corbett 87.

210 In *The History of Sexuality* Foucault shows how eighteenth-century regulations of sex became "anchorage points for the different varieties of racism in the nineteenth and twentieth centuries" (26), as sexuality became a central aspect of the self-affirmation of those in power.

211 Various scholars have shown this. See, for instance, Mahlis, King, Ferguson (*Where the Land Meets the Body*), or MacDonald-Smythe ("Authorizing the Slut"). See also my discussion of the mother figure in "Biography of a Dress" as acting as an agent of colonial ideals (chapter 2.3, section "A Yellow Dress and a Brown Girl.")

her that my whole upbringing had been devoted to preventing me from becoming a slut; I then gave a brief description of my personal life, offering each detail as evidence that my upbringing had been a failure and that, in fact, life as a slut was very enjoyable, thank you very much" (127–128). Through her sexual behavior, Lucy thus seeks to free herself from persistent colonial claims on her body.[212]

The following analyses in this chapter attend to passages in Kincaid's works in which the female protagonists' bodies engage with colonial history and its contemporary continuations. In two sections, I examine how the texts negotiate and counter sexual violence and exploitation through performative reenactments of violations of the colonized female body. The first section on "The Knowledge of Resistance" considers abortion in the contexts of sexual freedom. While the knowledge of how to end unwanted pregnancies simply aids Lucy in enjoying her sexual freedom without worries, the later novel *The Autobiography of My Mother* frames abortion in the contexts of colonial history. Here, the refusal to bear children highlights the biopolitics of slavery and the nonparticipation in it as a rebellious act. In the second section on "Stagings of Sex," I consider the performativity of sexuality that is exhibited in the sadomasochistic sexual practices in *The Autobiography of My Mother.* Sexual encounters reenact colonial fantasies of regulating and subjugating female bodies, highlighting their scriptedness, subverting binary power structures, and enabling the protagonist Xuela to exert sexual agency by dominating her partner, who takes on the role of the white colonizer.

My thoughts on bodies engaging with history are guided by Judith Butler's notion of the performative. Butler makes a rhetoric of theater productive to think about bodily doing that participates in the construction of gender: "[G]ender is instituted through the stylization of the body and, hence, must be understood as the mundane way in which bodily gestures, movements, and enactments of various kinds constitute the illusion of an abiding gendered self," Butler writes ("Performative Acts" 519). In analogy, I read the constitution of colonized female bodies in Kincaid's text through their (re)enactments of the past.

> The body is not passively scripted with cultural codes, as if it were a lifeless recipient of wholly pre-given cultural relations. But neither do embodied selves pre-exist the cultural conventions which essentially signify bodies. Actors are always already on the stage, within the

---

212 The sections that follow investigate explicit bodily and sexual doing in terms of abortion and S/M practices. In this, I here focus on representations and reenactments of physical violence as historically perpetrated on colonized female bodies. A discussion of Kincaid's protagonists as "sluts" and their sexual performances challenging the sexual morality at the heart of this invective would rather attend to a negotiation of more explicitly discursive claims on female bodies. This work has convincingly been accomplished by other scholars, see especially Gary Holcomb or Antonia MacDonald-Smythe ("Authorizing the Slut").

> terms of the performance. Just as a script may be enacted in various ways, and just as the play requires both text and interpretation, so the gendered body acts its part in a culturally restricted corporeal space and enacts interpretations within the confines of already existing directives. (Butler, "Performative Acts" 526)

Through the "stylized repetition of acts" ("Performative Acts" 519), Butler further contends, identity is instituted, but it also bears the possibility to reinterpret or rewrite cultural scripts. It is the focus on repetition with a difference that makes Butler's approach to bodies and the performative productive in the context of my study, as I will concentrate on repetitions and reenactments of scenes of sexual violence in Kincaid's texts, to read them as symbolic repossessions of the female body that simultaneously rework and reinterpret colonial scripts of female sexuality and reproduction.

The Knowledge of Resistance: Abortive Practices

Several of Jamaica Kincaid's texts explicitly mention terminations of pregnancies. The knowledge of how to end an unwanted pregnancy is presented as part of womanhood in Kincaid's works. As much as the texts' female characters exist in a largely patriarchal society, the choice to bear children nonetheless is unquestioningly theirs. This section on abortive practices in Kincaid's texts explores the resistances to the colonization of women's bodies through their procreative properties. Kincaid's female characters' decisions to (not) participate in biopolitics through reproduction highlights how class, race, and history are negotiated on and through human female bodies.

The recurring references to abortion throughout Kincaid's oeuvre suggest that it may be as common in women's lives as menstruation.[213] Abortion is already mentioned in one of Kincaid's earliest works, the prose text "Girl" (first published in 1978 in *The New Yorker*, then collected in *At the Bottom of the River*). The short piece consists of a list of how to behave in a "female" way. As such, the mother's instructions are a manual on how to perform gender, when they inform the eponymous "Girl" how to be *a* "girl." Judith Butler emphasizes the performative aspect of female-ness: "To be female is [...] a facticity which has no meaning, but to be a woman is to have to *become* a woman, to compel the body to conform to a historical idea of 'woman,' to induce the body to become a cultural sign, to materialize oneself in obedience to a historically delimited possibility, and to do this as a sustained and repeated corporeal project" ("Performative Acts" 522, emphasis in orig-

---

213 See "Girl" in *At the Bottom of the River* 5, *Lucy* 69, *ABM* esp. 82, *My Brother* 174, *MP* 136.

inal). The list in "Girl" gives instructions for gendered domestic chores, such as cooking and cleaning skills, and for proper deportment, such as how to eat in front of others or how to smile. The succession of imperatives interweaves any rules a female body might follow in a conservative notion in order to establish "girl-ness," including which kind of labor it might perform or how it might interact with other bodies (male or female). The mother thus provides the script – to use Butler's term – that underlies the bodily performance of gender. She intends to initiate her daughter into the performances of gender, "an act that has been going on before one [here: both the girl and the mother] arrived on the scene. Hence, gender is an act which has been rehearsed, much as a script survives the particular actors who made use of it, but which requires individual actors in order to be actualized and reproduced as reality once again" (Butler, "Performative Acts" 526). The mother's litany of instructions and imperatives puts forth aspects of gendered performance in a list-like fashion, stringing its contents together through continuation and semicolons rather than through narrative cohesion. The resulting equivalence of listed aspects indicates the inseparability of the laboring and the sexualized body in gendered acts. The instructions here are both delimiting and empowering in their transfer of knowledge.

Among the different topics covered, the reader finds mention of how to realize the termination of a pregnancy:

> this is how to make bread pudding; this is how to make doukona; this is how to make pepper pot; this is how to make a good medicine for a cold; this is how to make a good medicine to throw away a child before it even becomes a child; this is how to catch a fish; this is how to throw back a fish you don't like. (Kincaid, *Bottom* 5)

I suggest that this reference to abortion in Kincaid's earlier work establishes the topic's position in her oeuvre. Mentioned in passing, almost hidden between instructions on food preparation, the mother's voice shares information on abortive medicine with her daughter. It is the ordinariness of the instruction that makes it so noteworthy here. Knowing how to free oneself of possible unwanted children is as mundane as making bread pudding or as catching fish.

My reading of the reference to abortion in "Girl" as sharing empowerment and female bodily autonomy disagrees with Diane Simmons's: she interprets the advice on abortion as an indicator for the instability of a reality marked by traumatic loss. Simmons writes that "in the last third of 'Girl' the mother's voice continues the litany of domestic instruction, but added now is comment on a frighteningly contradictory world, one in which nothing is ever what it seems to be. The continued tone of motherly advice at first works to lighten the sinister nature of the information imparted and then, paradoxically, seems to make these disclosures even more frightening; eventually we see that, in a world in which a recipe for stew slides into a recipe for the death of a child, nothing is safe" ("Rhythm" 468). While Simmons reads the ordinariness of abortion as nightmarish and frightening, I rather understand it as building resilience to the possible nightmare of unwanted motherhood in the face of everyday oppression by patriarchal societal structures. At the same time, such instructions on how to maintain a level of independence from possible claims on the girl's body are contrasted with three explicit warnings against becoming a "slut," which takes up rather misogynistic notions of female sexuality that are contrary to those of a self-determined female body.[214] I maintain that this contradiction is motivated by the complex characteristics of the texts' mother figures, who often simultaneously perpetuate patriarchal regulations of female bodies and provide their daughters with the knowledge and mindset to resist such claims on their bodies.

In *Lucy*, these two sides of motherly instruction are evident in its education about the female reproductive body. While the narrator Lucy foremost presents her mother as a distributor of colonial ideals who also projects them onto her daughter's body, this clear distinction of colonizing mother and rebellious daughter is challenged by the mother's navigation of a simultaneous allegiance to and a revolt against colonialism in sharing her knowledge of female biology and plant-based abortifacients with her daughter. In contrast to Annie John's experience with menstruation during puberty, which is linked to her separation from her mother, in Lucy's case the onset of menstruation leads to a strengthened mother-daughter bond based on a shared bodily function. Lucy's mother is kind and reassuring at her daughter's fright of the bleeding. Here, menstruation does not lead to the experience of alienation, but to a feeling of community. More importantly, this female bonding is further expanded when the mother includes her daughter in a genealogy of knowledge of the female reproductive system. In this context, female sexuality is still not explicitly verbalized, but it presumes it is possible to imagine women freely partaking in sexual activity. Much later, when Lucy indeed has sexual relations, the information gained from her mother enables her not to worry about unwanted pregnancy:

---

214 Moira Ferguson also reads the mother's warning as contextualizing them with colonial history: "If only through discursive repetition, the daughter is obliquely reminded that 'slut' as a cultural taboo marks colonialism. Its overuse – in a quiet reference to the historical sexual abuse of black women – underscores the status of the narrator as a young black woman. The mother's admonitions illumine the critical role that history has played and is playing in everyone's lives" (*Where the Land Meets the Body* 15).

> If it [menstruation] did not show up, there was no question in my mind that I would force it to do so. In knew how to do this. Without telling me exactly how I might miss a menstrual cycle, my mother had shown me which herbs to pick and boil, and what time of day to drink the potion they produced, to bring on a reluctant period. She had presented the whole idea to me as a way to strengthen the womb, but underneath we both knew that a weak womb was not the cause of a missed period. (*Lucy* 69)

The topics of sexuality and the workings of conception remain silent in this exchange as Lucy recalls it. Still, knowledge is communicated through the joint picking of herbs and potion-boiling. Lucy and her mother partake in what Diana Taylor terms "the repertoire." Taylor differentiates between "archive" and "repertoire" to describe two realms of memorial practices. Taylor describes the well-established memorial medium of text as the "archive": "'Archival' memory exists as documents, maps, literary texts, letters, archeological remains, bones, videos, films, CDs, all those items supposedly resistant to change" (19). To consider social and cultural memory not included in such records, Taylor suggests looking to "performed, embodied, behaviors," which might uncover the stories and memories yet marginalized by the archive (xviii). Taylor proposes to take into account the "repertoire," which for him considers performance as an epistemological and memorial practice:[215]

> The repertoire [...] enacts embodied memory: performances, gestures, orality, movement, dance, singing – in short, all those acts usually thought of as ephemeral, nonreproducible knowledge. [...] The repertoire requires presence: people participate in the production and reproduction of knowledge by "being there," being a part of the transmission. As opposed to the supposedly stable objects in the archive, the actions that are the repertoire do not remain the same. The repertoire keeps and transforms choreographies of meaning. (Taylor 20)

The shared activity of mother and daughter in *Lucy* transfers knowledge from one generation to another. The repertoire, which operates through the body as a medium of knowledge, provides Lucy with the agency to govern her own body; she may "force" its functions, "reluctant" as they may be.

The narrative of Lucy is set in a North American city in the second half of the twentieth century. The text details the sexual liberation of a young woman who regards sexuality, contraception, and abortion as her own choices. Lucy's knowledge of abortive practices might thus be read as the personal version of a sexual

215 Taylor's use of the term "performance" here encompasses a wide range of physical doing, such as "[c]ivic obedience, resistance, citizenship, gender, ethnicity, and sexual identity, [which] are rehearsed and performed daily in the public sphere" (3). In the widest sense, what Taylor refers to here can be grasped as "expressive behavior" (xvi) or as "[e]mbodied practice" (3) which participates in the making of memory and knowledge.

revolution. The young woman's access to a means of birth control liberates her physically, but also from the claims of sexual morality, i.e., from patriarchal definition of womanhood regarding all aspects of life. In this, Lucy's knowledge of how to force menstruation may be read alongside the freedoms gained by women when the contraceptive pill was released to the marked – in the United States, the Food and Drug Administration approved "the pill" in 1960.

The Autobiography of My Mother is set much earlier in the twentieth century (specific years cannot be pinpointed) on the island of Dominica. Hence, this narrative is set in a much more specifically colonial space than *Lucy. The Autobiography of My Mother*, then, is more explicitly concerned with the history of colonialism, slavery, and the dehumanizing effects of the plantation system on people – including after the abolition of slavery. Accordingly, this later novel contextualizes abortive practices historically with the exploitation of female reproductive bodies during slavery and colonialism. As I will show, the abortions of Xuela Claudette Richardson recall the terminations of pregnancies that may have been performed by enslaved women before the abolition of racial slavery in the Caribbean.[216]

At age fifteen, Xuela Claudette Richardson is sent from her father's house to Dominica's capital Roseau to live with a work acquaintance of her father's. Xuela is to carry out domestic duties in the household of Monsieur Jacques and Madame Lise LaBatte for room and board. Staying at Roseau allows Xuela to further attend school, as her father wishes, and to study to perhaps eventually become a schoolteacher. Xuela's comment on this housing arrangement already hints at the sexual services she is to perform: "I was a boarder, but I paid in my own way. In exchange for my room and board in this house I performed some household tasks. I did not object, I could not object, I did not want to object, I did not know then how to object openly" (*ABM* 63). Xuela here does not explicitly mention the sexual relationship she ultimately has with Jacques LaBatte, yet the objectionable in this quote insinuates it. To Xuela, sex with Monsieur LaBatte is not necessarily desirable, but as she points out, disagreement cannot even be considered as an option. Rather, serving the owner of the home is presented as mundane, like a common domestic chore. Simultaneously, Xuela's comment here emphasizes her powerlessness as a very young woman in a likely distressing situation.[217]

After her arrival in Roseau, Xuela initially is content with her new housing arrangement, working for the LaBattes, attending school, and developing a deep

216 Slavery was abolished in the British colony of Dominica in 1834.

217 This power dynamic changes in the narration of the concrete sexual act, as the following section on the "Stagings of Sex" will present.

friendship with Madame LaBatte, whom Xuela pities for her loneliness, as she is married to a largely indifferent husband and without children. Still, the ostensibly warm and caring female companionship of Lise LaBatte and Xuela proves treacherous to the girl, as it lays the foundation for Xuela's sexual exploitation. In intimate moments with her motherly friend, Xuela anticipates the requests the LaBattes will eventually – wordlessly – make of her. "[S]he wanted something from me, I could tell that [...]. It never crossed my mind that I would refuse her. She wants to make a gift of me to her husband; she wants to give me to him, she hopes I do not mind" (*ABM* 68). Interestingly, this exchange regarding sexuality passes without words. As Nicole Matos notes in her careful study of cloth in *The Autobiography of My Mother*, this intimate communication between the two women is facilitated by a dress, once worn by Lise LaBatte, now given to Xuela. The dress becomes a medium that transports Madame LaBatte's thoughts and intentions to the girl (see Matos, "The Difference" 848). Hence, in sharing desires about sexuality, Lise and Xuela communicate via shared doing, rather than words – similar to how Lucy is introduced to knowledge of abortifacients by her mother.

The first sexual encounter with Jacques LaBatte then is called "inevitable" by Xuela (*ABM* 69). The ensuing frequent sexual acts between employer and servant girl are presented in a double-bind of coercion and enjoyment, which will be at the center of the following section on the performativity of sex in Kincaid's works. First, however, the analysis will focus on the resulting pregnancy and its historical implications.

I emphasize Madame LaBatte's involvement in her boarding servant's sexual relationship with her husband to highlight that Xuela's first pregnancy, which resulted from the nighttime meetings with Jacques LaBatte, is part of a larger structure of exploitation that extends well beyond the physical encounters of two people. Eventually it becomes clear to Xuela that Lise LaBatte's wish that Xuela sleep with her husband is motivated by her own desire for a child. Incapable of producing a baby with her own body, Lise wants her servant girl to bear her husband's offspring (*ABM* 77). In her own "defeat," as Xuela immediately recognizes Lise's subject position (*ABM* 65), Madame LaBatte seeks to use the body of another woman to establish her own worth in a patriarchal society that values women expressly for their reproductive capacities. "The reason why children are so important for women can be partly attributed to the fact that, in a world of inequalities where stability is fragile, women's survival depends largely on the offspring they can bear, thus making their own bodies extensions of those of men through their children," as Guillermina De Ferrari emphasizes (171). Accepting and perpetuating this subjugated position that is rooted in reproductive capability, Lise LaBatte collaborates with the discursive power of patriarchy by actively extending it to Xuela's body.[218]

Structurally, Xuela's relationship and living arrangements with the LaBattes recalls the patterns of power of the plantation: the girl is left with her new masters to serve them in all domestic aspects. She enters the stately master's home only to be of service, while her own living quarters are relegated to the back of the property in a rougher shed next to the kitchen. Xuela is utterly powerless, unable to govern her own body so that even the thought of rejecting sexual advances is inconceivable. In their relationship with Xuela and with their claims on her body to please them sexually and reproductively, the LaBattes take on the roles of colonizers. Particularly the crucial aspect of sexuality and pregnancy in their interactions marks them as such. Orlando Patterson shows that the exploitation of female sexuality was an intrinsic aspect of slavery (173). Women were exploited for their bodies as working units both in terms of hard labor and in terms of reproductive labor to increase the colonizers' prosperity. In her study of the role of female enslavement in transatlantic slavery, Jennifer Morgan pointedly notes that "[w]omen's labor was at the heart of monoculture export economies in both the Caribbean and the American South, and their reproductive lives were at the heart of the entire venture of racial slavery" (4). Although *The Autobiography of My Mother* does not identify the LaBattes as white or distinctly French,[219] given their repetition of colonial structures it is unsurprising that some scholars indeed read Jacques and Lise LaBatte as white (for instance Alexander, "Postcolonial Hauntings" 125).

Having conceived a child from Monsieur LaBatte under these circumstances, Xuela decides to abort the pregnancy and seeks help from an older woman – most likely an herbalist – who prepares a strong dark drink for the girl and provides her with the space to suffer the abortion. Again, this proceeds wordlessly (*ABM* 82). Xuela lies in pain for four days until she bleeds from her womb.

---

218 This collaboration is also indicated by Lise's change of language upon learning of Xuela's pregnancy, as Giselle Anatol notes: while Lise and Xuela generally converse in French patois, a language which Xuela had earlier denoted as "the language of the captive, the illegitimate" (*ABM* 74), Lise now switches to English, "the formal language of oppression […]. The use of English, the so-called 'father-tongue,' marks her [Lise's] complicity with patriarchal dictates for the oppression" (Anatol 946).

219 The text indeed does not mark them racially, for instance in terms of skin color, at all. In this novel, Blackness generally is unmarked while deviation from African heritage is commented on, such as is the case with Xuela's Carib mother, her half-Scot father, or her later white British husband Philip. Hence, I rather read the LaBattes as Black Dominicans. (Rhonda Cobham and Guillermina De Ferrari arrive at the same conclusion.)

After this immediate ordeal, the pain in her lower body and the bleeding continue for an unspecified amount of time (*ABM* 91).

Using folk medicine and plant-based abortifacients has a tradition going back to and beyond the times of slavery in the Caribbean. During the late eighteenth and early nineteenth century, birth rates among enslaved women in the Caribbean were strikingly low, baffling slave holders at the turn of the century. Perhaps unsurprisingly, interest in women's reproductive bodies developed along the restrictions of the import of new slaves (see Schiebinger 142 or Bush-Slimani 86–87). Slave holders and planters at the time attributed low fertility to the harsh working and living conditions of enslaved women, but also to what they determined as immorality, promiscuity, or idling (see Bush-Slimani 88–89 or Schiebinger 143–145). Indeed, birth rates were so low that without a steady influx of new enslaved people, Caribbean enslaved populations would have decreased.[220] Among such reasons as amenorrhea (absence of menstruation) for physical and psychological stress, the attempt to avoid sexual intercourse as not to participate in breeding capital for enslavers, abortion must be counted as a cause for the low birth rates. Historical records show that planters and slave holders counted knowledge of abortive practices among enslaved women as a factor in low fertility, as for instance Bush-Slimani, Schiebinger, or Morgan show. "Historians today have no way of saying how many women killed their unborn children in the West Indies. That abortion was practiced, is not in doubt, however," writes Londa Schiebinger in her study on natural abortifacients (149). Barbara Bush-Slimani attributes contemporary absence of knowledge regarding slave reproduction to a historical blind spot based on gendered disbalances in records: "Absence of hard evidence should not be taken as proof that the practices [of abortion] did not exist. Because of the strong taboos against abortion in most patriarchal societies, knowledge of abortion practices has always been in the shadowy and hidden world of women outside the orbit of mainstream history" (93).[221] In this context, we may also read the silences regarding abortion in Jamaica Kincaid's works. As, for instance, also seen in the

---

220 See Schiebinger's work, which presents estimations by planters and historians (143) and, for instance, the calculation that one in 4.6 pregnancies was not carried to term, if women conceived at all (130).

221 We might read the elusive suspicions of eighteenth-century colonists as illustrating this knowledge barrier, as well as the late emergence of historians' interest in the topic. Considering actively induced abortion as a possible strategy of resistance zooms in on the positions of women in enslavement, which in historical scholarship was long marginalized, as for instance Bush or Morgan highlight: The particularities of female enslavement were largely neglected by historians, esp. before the 1980s (Bush 121). Morgan still criticizes the lack of scholarship on reproduction and enslaved women's roles in it in 2004 (6).

interactions of Lucy and her mother mentioned earlier, knowledge of the female body is repeatedly transferred though shared activity. Bodily doing and objects, such as Lise LaBatte's dress, become media of communication. It is moreover noteworthy that while all of this is transported by text in Kincaid's works, again these texts remain silent when it comes to specifics. No recipes or instructions on how to actually prepare abortive medicine are shared with readers. The characters are shown to exchange them, but the recipients of Kincaid's texts remain on the outside of this community.

Londa Schiebinger, in her study of plants and empire, focuses especially on the "peacock flower," *flos pavionis*. Schiebinger investigates the nontransfer of knowledge by the example of this plant, which, as she shows, was known for its abortive properties to both indigenous and Afro-Caribbean populations (specifically Arawak and enslaved African women), while Europeans largely remained ignorant to the plant's medicinal use (Schiebinger 2). If we read the mentions of plant-based abortifacients in Kincaid's works alongside such intermittent structures of knowledge transfer, it becomes clear that the texts perform the same strategy of inclusion and exclusion – especially if we keep in mind that the readership of Kincaid's American-based publisher likely largely consists of people of European descent. In this, these texts simultaneously highlight the existence of women's knowledge and its marginalization and exclusion from verbal discourse.

Considering the external rule enslavers enforced on all aspects of enslaved women's lives, their refusal to bear children to the plantation system certainly constitutes acts of resistance. Women's bodies and sexuality thus became a ground of political negotiation (see Schiebinger 149 or Bush 121). Xuela's abortion in *The Autobiography of My Mother* suggests itself as a resistance to the commodification of her body as a sexual object and breeding vessel in the same way.[222]

Indeed, Xuela explicitly expresses her desire to be free of the claims laid on her by having a child (*ABM* 89), as this would mean for her to give up on herself, on her own future and desires (*ABM* 82). Ridding herself of the pregnancy, Xuela emerges as "a new person" (*ABM* 83). Through the abortion of her pregnancy, as Simone Alexander puts it, "Xuela rejects her designation as propertied possession" ("Postcolonial Hauntings" 116) and she "reconstructs the black female body as a site of resistance and resilience" (Alexander, "Postcolonial Hauntings" 117). While I agree with Alexander that Xuela's negotiation of subjecthood is carried out on and through her body, I also emphasize that it does not come away from this un-

222 As such, and as withholding her child from being owned by others, *The Autobiography of My Mother* here is reminiscent of Sethe's killing of her child Beloved in Toni Morrison's treatment of childbearing and rearing during slavery in her landmark novel *Beloved*.

harmed. Further violence and pain are inflicted on Xuela's body, but with the important difference that now the choice to do so was Xuela's own.

Alexander and De Ferrari then read Xuela's emergence from her ordeal as a "rebirth" ("Postcolonial Hauntings" 124, De Ferrari 151). However, Jennifer Morgan reminds us that a "reductive view of reproduction does not move us beyond binaries" (167). The decisions of enslaved women to abort their pregnancies was not their only means of resistance,[223] nor might resistance have been their only reason to abort.[224] Rather, Morgan calls attention to the complexities of reproduction during slavery. Morgan's cautioning might also be taken to heart in reading Xuela's abortion as a repetition of the reproductive choices Caribbean enslaved women likely made. While I agree with other scholars who read Xuela's choice to end her unwanted pregnancy as empowering, I emphasize that this "rebirth" is a slow process and that the power to be gained by Xuela initially consumes her herself. Xuela becoming a new person also means that she has to become undone first: upon realizing that she indeed is pregnant, Xuela weighs future and past, presence and absence, her own existence and nothingness:

> I believed that I would die, and perhaps because I no longer had a future, I began to want one very much. But what such a thing could be for me I did not know, for I was standing in a black hole. The other alternative was another black hole. This other black hole was one I did not know; I chose the one that I did not know. (*ABM* 82)

The Autobiography of My Mother repeatedly evokes bleakness and darkness as a general condition. For instance, the death of Xuela's mother and the girl's orphanhood are described in these terms in the very first sentences of the novel: "there was nothing standing between myself and eternity; at my back was always a bleak, black wind" (*ABM* 3). It seems obvious to read the black hole Xuela always already inhabits as a metaphorical reference to the nothingness of History that affects Xuela. Her female genealogical link to the past was broken twice, as not only

---

223 Withholding the reproductive capacities from the economic system of slavery was but one way in which enslaved women resisted. See, for instance, Barbara Bush, *Slave Women in Caribbean Society 1650–1838*, esp. 51–82; or Jennifer Morgan, *Laboring Women*, esp. 176–195, for further investigations of the resilience of enslaved women in the Caribbean. Both scholars highlight that considering only women's bodies in this context would yet again be a reduction both of the complexities of their lives and their resourcefulness.

224 Morgan points out that also childbirth might have been a means of resistance during slavery, as "[t]hrough the birth of their children, enslaved women may have seen a means to reappropriate what should have been theirs all along" (132). Still, also this position should not be reduced to the image of the sacrificing slave mother, which in the ideology of perfect domesticity restores the woman's humanity, yet reduces her to the role of the doting maternal caregiver (Morgan 167).

she, but also her mother, was orphaned; she is part of a group of people regarded as eradicated and extinct, i. e., her Carib heritage is marked by absence. Communication with the past is thus silenced twice over, both personally and collectively, and the texts of History do not provide space for her either. Reading the metaphors of the two black holes in context with Xuela's abortion emphasizes that either decision has its downfalls. Neither carrying the child to term nor upending the pregnancy before that are options that offer happiness, but both are different paths in darkness.

Finally, Xuela's reference to a black hole here is also a literal one. She is led to lie in "a small hole in a dirt floor to lie down" by the herbalist who assists Xuela with the abortion (*ABM* 82). Eventually, Xuela rises again from this hole. Alexander interprets this as leaving a womb ("Postcolonial Hauntings" 121), while I propose to read it as a shallow grave. Regarding the abortion as an empowering act for the young woman and as a repetition of historical resistance, this literal hole is superimposed with the black hole of history from which Xuela attempts to free herself. Yet I maintain that this is far from simply leaving the shelter the old woman had provided and entering a new self-determined life, as the term "rebirth" might suggest. Rather, Xuela stays in such a metaphorical hole of nothingness for quite a while. After healing from her physical wounds at her father's house, she returns to the LaBattes, where she is met with Madame LaBatte's grief over the loss of the potential baby that she had wanted for herself. Wishing to leave her own grief and the memory of her ordeal behind and unable to create a distance to it in time, Xuela creates a distance in space by leaving the LaBattes.

Xuela's departure comes with a change of outward appearance. She thus processes her experience once again physically. She buys "from a wife the garments of a man who had just died: his old nankeen drawers, his one old pair of khaki pants, his old shirt of some kind of cotton" (*ABM* 98). Xuela also cuts her plaits and "wrapped [her] almost hairless head in a piece of cloth" (*ABM* 99). Moreover, Xuela leaves the domestic space and with it a gendered form of labor commonly regarded as feminine, in order to become part of a work crew that builds a road. She now takes on physically much more taxing labor outside the home.

Ostensibly, with these changes in appearance and occupation, Xuela toys with notions of gender – exchanging the typically feminine for a performance of masculinity. However, Xuela's cross-dressing here is not a playful negotiation of gender roles at all, but an expression of loss. While she indeed "adopt[s] different social identities" through cloth, as Anne McClintock in brief sums up cross-dressing (*Imperial Leather* 173), Xuela's donning a man's outfit is less geared towards trying on a male social role than towards disappearing from societal claims on her by not identifying with either a male or female role in this change of clothing. "I did not look like a man, I did not look like a woman," Xuela notes (*ABM* 99). As Marjorie Garber shows, cross-dressing permeates the seemingly firm borderline between binarities, such as Black/white, master/servant, male/female (16). As such, Xuela's cross-dressing and her change in occupation plays on the ambiguities between the binary pairs and allows her to inhabit a space of nondefinition.

Even more than that, Xuela finds herself "living in this way for a very long time, not a man, not a woman, not anything" (*ABM* 102). In her own view, Xuela thus leaves the realm of binarities and enters a space of undefined ambiguity. Xuela refuses to perform the gendered scripts of femininity. Her taking on male props (clothing) still does not propel her into a performance of masculinity. She stops short in an in-betweenness and remains in this state of undecidability, effectively withholding herself from the structures of human interaction around her. Reading this with a somewhat hopeful notion, Kathryn Morris points out that "[t]he androgyny and her seclusion also allow a moment of transcendence from the realm of the social order where she exists as a woman, a fated, gendered subject" (964). Still, if we read Xuela's sexual exploitation and possible childbearing body in analogy to the lives of enslaved women, this offers another possible interpretation of the de-gendering of Xuela's body. Bush-Slimani for instance shows the doubled conceptions of the female slave body in its productive capacity both as an asexual labor unit and as unit of biological reproduction (84). Xuela's attempted nonperformance of gender in conjunction with taking on hard labor thus repeats the other role of enslaved women – that of the supposedly ungendered laborer.

More important than the masculinity of the clothes in which Xuela now clads herself, is their origin as having belonged to a dead man. Xuela inhabits a space of in-betweenness, not only in terms of gender, but also in terms of life and death. She has not allowed a life to form inside herself and to expel it, and she underwent a near-death-experience. Afterwards, she does not directly return to life, but in a traumatic repetition remains in this interstice of nothingness. Every morning, she drinks a strong cup of coffee, a "beverage that was thick and black and […] pungent in flavor" (*ABM* 98). The repeated ingestion of this strong black drink points to a symbolic repetition of the consumption of the "thick black syrup" that initially had induced the abortion (*ABM* 82). Xuela's life now is enveloped by death. Absconding motherhood once and for all, she imagines the powerful, god-like destruction of any future children in downright biblical terms of God's wrath, such as unquenchable thirst or distorting disease (*ABM* 97–98). Kathryn Morris reads Xuela's imagined consumption of her possible future offspring in terms of what she calls the "Caribbean literary romance with death [which] might speak to the performance of the trope of cannibalism which metaphorizes the physical ingestion of flesh into a nostalgic incorporation of a lost body into a present body" (956). I concur with this in that I also read Xuela's excessive imagined acts of power, perpetrated on unprotected bodies, as an expression of anger and grief over her former powerlessness and loss. She thus gains power over what she forced not to be part of herself – a child that she did not want to exist. Xuela's abortion indeed is empowering to her, but power comes with loss. Xuela repeats the severing of genealogical lines, just as she is separated from her mother, and her mother from her mother before that. In this way, Xuela links herself to her heritage of absence. In the absence of others, Xuela learns to love herself. She fills the "nothing" inside of her (*ABM* 99) with herself. In classic narcissistic fashion, she falls in love with her own picture, beholding herself in a pool of water and finding pleasure in her own body (*ABM* 99–100).

Xuela eventually leaves her self-imposed reclusion and returns to both wearing women's clothing and to living with her family. For the time being, her self-actualization is complete. She leaves the state of nothingness and ghostliness by burning her male clothes and departing in the middle of the night so as not to bring attention to her absence (*ABM* 104) as she returns to being a presence. Her distinctive self-love will stay with her.

Ultimately, Xuela's abortion highlights the repercussions of slavery and colonization on family structures, as Xuela repeats the breaking of genealogical links. The novel examines the absence of female genealogy, similar to the later book *Mr. Potter* attending to the absence of familial bonds in a male genealogy. Yet, the important difference in Xuela's repetition of abortive practices is her successful reclaiming of her own body by doing so. While she suffers, she has the power to govern this suffering as she pleases.

The Autobiography of My Mother replaces the absence of personal genealogy with that of women's shared bodily experience. While Xuela is dispossessed of a personal history, she is thereby linked with a Caribbean female collective history. Where personal memory must fail, the repertoire of plant knowledge establishes a female community with roots in a shared history that counters the black hole of History with a capital *H*. In this genealogy of shared knowledge, Xuela importantly draws on the history of both Afro-Caribbean and Indigenous women. While the expertise of plant-based abortifacients is recognized as traditional West African knowledge, often ceded in female community from mother to daughter (Bush-Slimani 90, 141), it also was part of Indigenous life in the Caribbean (esp. of Tainos, Caribs, and Arawaks) before European colonization (Schiebinger 128). Though it is now impossible to exactly trace how knowledge may have been shared between enslaved Africans and Indigenous peoples, Schiebinger shows overlaps in familiarity with endemic abortifacients, such as the peacock flower (2). Creole Caribbean cultures thus incorporated both West African medicinal expertise and local knowledge (see also O'Donnell 102).

Xuela in Kincaid's novel frequently laments the extinction of her Carib ancestors, yet with her herbal abortion, she can take up the traditional teachings of her forebears. Sadly, this presence of historical knowledge yet again reinforces and emphasizes the absence of Indigeneity here, as with Xuela's pregnancy it also aborts Carib future. This once more highlights Xuela's inheritance of nothingness. Veronica Gregg also notes that Xuela's abortion replicates "the aborted life that is her inheritance" (928). Simultaneously, because of its self-making potential, it offers a way to conquer this history of brutality and death. As such, "Xuela's abortions stand as a figure of the problems of West Indian language and history," Gregg determines (929). I add that with its circularity and revising repetitions, Xuela's abortion notably takes up the structures in which, as shown throughout this study, Jamaica Kincaid's works productively engage with both personal and collective history. What stands out here is Xuela's stage to negotiate history – her own body. The perpetration of sexual and other exploitative violence is inflicted on her, pointing to the long history of abuse of Black women's bodies. Yet, it is also via her body and via shared bodily doing that the text counters the erasure of women's bodies from history. As such violence physically disrupts female genealogies, in *The Autobiography of My Mother* shared women's history is moved from physical to cultural continuance. This collective knowledge – which pertains to female bodies – constitutes a female community that as repertoire in Taylor's sense withstands the violence of discursive History, which largely marginalizes them. This community is now not based on family ties but on a shared body of knowledge.

Stagings of Sex: S/M and Colonial Fantasy

Bodies in Jamaica Kincaid's works not only counter and defy discourse, but also take up memory work in moments where verbal communication is either absent or less usable for the texts' characters. This second section analyzes how the text presents concrete erotic or sexual acts and how this contributes to the novels' discussions of the entanglements of gender, race, and history. It is noteworthy that many of the erotic relationships in Kincaid's novels are marked by a confluence of pleasure and pain. Acts of affection or erotic arousal are often coupled with physical violence. I read this double-bind through the lens of S/M conceptions in order to highlight the performative aspects of the sexual (or erotic) interactions of human bodies in three of Kincaid's novels – *Annie John*, *Lucy*, and *The Autobiography of My Mother*.

Practices of S/M (sadomasochism) are highly ritualized. Their focal point is explicitly *not* the usurping of power by one dominating party, exerting all-encompassing power on a submissive subject. Rather, as Anne McClintock shows, this assertion of power is part of ritualistic repetition and subversion of power structures. S/M revolves around an "economy of conversion: master to slave, adult to baby, power to submission, man to woman, pain to pleasure, human to animal and back again" (McClintock, *Imperial Leather* 143). How do human bodies in Kincaid's texts interact in meaningful ways that negotiate such binaries? How do bodies comment on the history of colonialism through erotic practices? Observing the performativity of S/M in this context opens up erotic and sexual scenes to readings of subversion of power instead to accepting the painful subjugations of the texts' protagonists. The structure of my analyses follows the chronology of the three novels. *Annie John* was not only published first; it also features the youngest protagonist. Erotic play here does not yet use the paraphernalia of S/M, nor does it negotiate colonial history as explicitly as does *The Autobiography of My Mother*, but a more personal power structure between mother and daughter is at stake. Nonetheless, this chapter begins with an examination of relevant passages from this early work to chart a trajectory in Kincaid's oeuvre, which builds up from *Annie John* and *Lucy* to two more explicit sexual scenes in *The Autobiography of My Mother*, one of which may be read as rape, which signifies historical violations of female bodies. The sexual relationship with Xuela's later husband then presents an overt example of the bodily enactment (and reenactment) of such power structures, simultaneously subverting them by staging them through S/M play.

In *Annie John*, sadomasochistic behavior occurs less in heterosexual relations as in homosocial (and arguably homosexual) scenes between young girls. The text details Annie John's growing up from childhood to young adulthood. Once she leaves her parents' house to attend school, she forms intense friendships with other girls. While these are the relationships of pubescent girls en route to discovering their sexual bodies, and their physical interactions appear playful, they are no less erotic than the physical relationships of the adult protagonists in the later novels *Lucy* and *The Autobiography of My Mother.*

Annie John's earliest friendships already feature a doubling of affection and cruelty. Annie, for instance, likes to buy a sweet treat for her friend Sonia, to then tug on the girl's body hair until she cries while she simultaneously enjoys Annie John's gift (*AJ* 7). "Love" and "torment" here occur in the same sentence and relationship (*AJ* 7). Declarations of love are accompanied with flurries of kisses (*AJ* 50), and body parts – including breasts – are presented to each other (*AJ* 52, 80) in a seeming competition of growth.

Both eroticism and S/M practices are then most pronounced in Annie John's relationship with the Red Girl. As discussed in the previous chapter on "The 'Unruly' Female Body," this girlfriend is of particular importance to Annie John because she transgresses boundaries – be it by disregarding gendered social roles or by refusing to adhere to hygienic norms. Annie John is charmed by this breaker of rules and attempts to woo her with presents. Meeting in secret by the lighthouse, the girls follow the scripts of a romantic relationship. Eventually, Annie is forced to stay away for a few days after having been found out by her mother, who disapproves of her daughter's contact with a social outsider. Annie John thus abandons the Red Girl who keeps waiting for her in their usual meeting place. Once Annie turns up again, their encounter proceeds wordlessly until it turns into a bodily performance of the girls' emotions:

> Then, still without a word, the Red Girl began to pinch me. She pinched hard, picking up pieces of my almost nonexistent flesh and twisting it around. At first, I vowed not to cry, but it went on for so long that tears I could not control streamed down my face. I cried so much that my chest began to heave, and then, as if my heaving chest caused her to have some pity on my, she stopped pinching and began to kiss me on the same spots where shortly before I had felt the pain of her pinch. Oh, the sensation was delicious – the combination of pinches and kisses. And so wonderful we found it that, almost every time we met, pinching by her, followed by tears from me, followed by kisses from her were the order of the day. I stopped wondering why all the girls whom I had mistreated and abandoned followed me around with looks of love and adoration on their faces. (*AJ* 63)

Annie John and the Red Girl here articulate themselves nonverbally. In the order of these events, the pinching might initially be read as a punishment for the abandonment of the Red Girl. In its combination with kissing and in the subsequently frequent repetition of this scene, however, the pain that the Red Girl inflicts on Annie John's body becomes a rehearsed performance of roles in their relationship.

Chinmoy Banerjee locates Annie John's physically painful practices with her girlfriend within S/M structures:[225] with the Red Girl, Annie John ostensibly performs the submissive part, passively being pinched and kissed. Her repeated giving of presents to the Red Girl similarly "establishes the dominant position of the Red Girl (basic to S/M games)" (Banerjee 43). Still, Banerjee maintains that the power positions do not settle into a "gendered hierarchy, thus staying out of traditional fixed positioning of S/M" (43).[226] Building on Banerjee's observation, I propose that the absence of a gendered (male-female) power structure here facilitates the Red Girl's appeal to Annie John. The denial of fixity Banerjee notes is the main attraction of the girls' S/M play here.

Annie John's submission to the Red Girl enables her to reevaluate other seemingly fixed power structures. Throughout the novel, Annie John seeks to establish independence from her mother, who attempts to regulate the girl's behavior and

---

225 Roberto Strongman also reads the pinching in *Annie John* (both with Sonia and the Red Girl) as a "coded form of sexuality" (31).

226 While I here agree with Banerjee, with McClintock I question whether S/M indeed "fixes" positions, as Banerjee seems to suggest. Instead, I will argue in the course of this chapter that S/M practices subvert binarities and fixity.

body. Her friendship with and submission to the Red Girl contrasts with the overbearing mother, who hitherto was the ruler of Annie John's life. Thus, Annie John's turn towards the Red Girl is a turn towards a new rule in her life that subverts the teachings of the mother. Kissing and pinching then are but one aspect of Annie's adoration of the Red Girl, her freedoms, and her challenging the power and fixity of her mother's rules.

Annie John's contact with the Red Girl enables further deviances: in order to bestow little gifts on her friend, Annie John steals money from her mother's purse (*AJ* 64). After having received three marbles from the Red Girl, Annie John is excited to spend her spare time playing marbles – an activity her mother had explicitly warned her against as unsuitable for girls. Annie John even perfects the game and to her satisfaction wins a prized marble for the Red Girl (*AJ* 65). For their secret meetings, the girls rendezvous at an old lighthouse, which is a place explicitly forbidden by the mother (*AJ* 58). Finally, the concealment of the friendship with the Red Girl leads Annie John to invent several lies and evasions to account for her absences. Striking up a secret relationship with her enables the girl to revolt against the mother's power over her in several small ways, such as lying, stealing, inappropriate activities and spaces, as well as absenting herself. In this way, submitting to the Red Girl's "un-rule" allows Annie John to transgress the boundaries that were previously set by her mother. Allowing a third into the tense mother-daughter connection, Annie John breaks with the binary she had been an integral part of and frees herself from the fixity of her mother's rules. As part of this, the physical S/M play with the Red Girl, which also transgresses binaries with its confluence of pain and pleasure, performs one aspect of Annie John's negotiations of power positions. With these subverting and liberating functions of S/M, *Annie John* sets the tone for further scenes that connect pain and pleasure in Kincaid's later works in ways that are sexually more overt.

The novel following *Annie John* continues to present eroticism and sexuality in and as negotiations of power. The protagonist of *Lucy* has left puberty behind and strives to establish her independence from her mother and mother country as a young female immigrant in the United States. Scenes of sexuality in this novel largely emphasize the protagonist's agency. For instance, reminiscing on her youth, Lucy reflects being assaulted at a library, when a distant acquaintance abruptly kisses her unsolicitedly. Lucy recalls that he "pressed his lips against mine, hard, so hard that it caused me to feel pain, as if he wanted to leave an imprint. I had two reactions at once, I liked it and I didn't like it" (*Lucy* 50). This perpetration on her body, which seeks to mark her as a possession of a random boy, as the term "imprint" suggests, does not leave Lucy violated and helpless. Instead, her reaction to the assault is doubled in pleasure and rejection. Lucy then immediately leaves the role of passivity and returns the favor by kissing the boy back. In this, she takes it a step further, involves her tongue and leaves the boy perturbed: "The whole thing was more than he had bargained for, and he had to carry his school-bag in such a way as to hide the mess in the front of his trousers" (*Lucy* 51). Whereas the hard kiss felt like an "imprint," i. e., a marking that affects a surface by indenting it, Lucy's tongue kiss penetrates the boy's body by invading his mouth. The boy's consequent involuntary ejaculation leaves the initial aggressor hiding and embarrassed. This episode presents the bargaining of sexual agency in erotic exchange, just as it establishes Lucy as aggressive and capable. Her sexual body is a tool she can use for both pleasure and power, it will remain so throughout the novel.[227]

In Lucy's later relationship with Paul, she enjoys her sexuality and emphasizes her own pleasure. She muses, "how surprised I was to be thrilled by the violence of it (for sometimes it was that, violent), what an adventure this part of my life had become, and how much I looked forward to it, because I had not known that such pleasure could exist and, what is more, be available to me" (113). Almost in passing, these lines refer to violent sexual practices. It remains unclear who – Lucy or Paul – might be the perpetrator or receiver of such violence, or whether the roles might be shifting interchangeably. It even remains unclear whether the violence indeed is physical, although this may be assumed based on the contexts. What is noteworthy, however, is the definite pleasure that another of Kincaid's protagonists derives from violent sexuality. Though not obviously defined as S/M practices, Lucy's comment here nonetheless emphasizes violence as part of sexual normality in her life. This might be kept in mind in reading Xuela's sexual preferences in *The Autobiography of My Mother*. As the above examples from *Lucy* show, Kincaid's oeuvre does not condemn physical violence in erotic encounters. Rather, this type of bodily interaction is framed as an enjoyable aspect of physical power negotiations.

Finally, the 1996 novel *The Autobiography of My Mother* presents the most overt cases of sadomasochistic sexual interaction in Kincaid's works. Hence, the following main part of this section will be devoted to the analysis of two of the sexual relationships that the protagonist Xuela maintains.[228] The violence in both allows for positions of domination and submission to be assumed, negotiated, and highlighted. In part, this performative highlighting also subverts seemingly rigid power positions, which I read as repetitions of colonial power structures.

---

227 See also Kristen Mahlis, who reads "Lucy's awareness of her body as a source of resistance and sexual pleasure" (165).

228 Xuela has sex with more than the two men discussed here. For instance, she sexually enjoys the affair with Roland, a stevedore. Yet the sexual relations with Monsieur LaBatte and her later husband Philip Bailey are most interesting in the context of S/M practices, which is why this section solely focuses on Xuela's sexual interactions with the two of them.

I will first focus on the sexual relations with Monsieur Jacques LaBatte, which led to the pregnancy and abortion discussed in the previous section. These first scenes of sexuality in *The Autobiography of My Mother* have variously been read as LaBatte raping Xuela.[229] Other scholars here read a somewhat willing participation on Xuela's part.[230] Given the exploitative structures in which Xuela is trapped,[231] interpreting Xuela's sexual encounter with Monsieur LaBatte as rape points to the systemic violence that confronts her. Carine Mardorossian, for instance, considers the historical implications of sexual abuse of women of color and contends that Kincaid's novel "exposed how rape has historically been configured as sexual aggression between classes and races and how victims and perpetrators have consistently been distinguished from one another on the basis of other categories of identity than gender" ("Rape by Proxy" 115). In other words, rape here does not signify the violation of a singular body, but with the historical conjunction of rape and race, issues beyond the personal are at stake. Nonetheless, I would here like to offer a slightly different reading, as a consideration of the details in the sex scenes such as space and lighting provide avenues to reading Xuela in these violent sexual encounters not necessarily as a passive victim opposite a forceful aggressor, but as an active participant in the events. My readings seek to enter a conversation with those of other scholars. While differing at times, I do not intend to downplay the violence of sexual aggression. Rather, I see my interpretations on the following pages as complementary to existing ones and emphasize the openness of the text to multiple, seemingly contrary readings.

Over the course of the four pages that detail the first time Xuela and Jacques LaBatte have sex, the text offers different interpretations – either as the exploitation of a girl in a vulnerable position, or as a consensual act, or a simultaneity of these two extremes. This variability, I argue, is a result of the contraction of pain and pleasure and how readers and scholars chose to evaluate the use of force in sexual acts. Xuela's initial description of her sexual encounter with Monsieur LaBatte as "inevitable" and her detailing that it "was no less a shock because it

229 See for instance Cobham, Mardorossian ("Rape by Proxy"), or Simone Alexander ("Postcolonial Hauntings").

230 See for instance Holcomb, Schultheis, or McDonald-Smythe ("Kwik! Kwak!"). It should be noted that in Antonia MacDonald-Smythe's 1998 article, a spelling error in the author's name occurred, dropping an "a" from her family name. For consistency's sake and so as not to confuse references, I have retained the erroneous spelling in my own bibliographical references to this particular article, but I call readers' attention to the correct spelling, which I will generally use when referring to this scholar.

231 See the previous section on abortive practices for a detailed analysis of Xuela's exploitation through the LaBattes, including a historical contextualization.

was inevitable" (*ABM* 69) already holds such ambivalence. Wherein exactly the inevitable or the shock lie, is left open to interpretation. In any case, this choice of words positions Xuela as vulnerable and the ensuing action is somewhat outside of her control. The same quote might also be read as a comment on the precarious position of women in a relationship of dependency with an employer, such as that of Xuela as a servant girl which leaves her open to sexual exploitation.[232] Then again, the same line might just as well refer to a female first experience of penetration of the vaginal canal. This transgression of a body's outside borders then incites the shock mentioned. In any case, the ambiguity of possible readings here superimposes female biology with gendered social precarity, emphasizing the vulnerable position of the young woman.

Xuela then continues her narrative of her first sexual intercourse with Monsieur LaBatte with his approach to her in the backyard of the LaBatte's property. Its textual details, I maintain, point to the complexities of the encounter – which is far from a simple opposition of powerful and powerless. Xuela is in the yard on her own at nightfall. She is enjoying touching herself intimately, when Monsieur LaBatte appears and wordlessly watches Xuela's masturbatory acts. This does not interrupt her, but she continues with her private enjoyment. "I saw that Monsieur LaBatte was standing not far off from me, looking at me. He did not move away in embarrassment and I, too, did not run away in embarrassment. We held each other's gaze" (*ABM* 70). Quite explicitly, Xuela here presents a state of equality in this exchange of gazes, rather than being overpowered by the male gaze. Xuela then brings her fingers to her face and enjoys the smell of her own genitals. "It was the end of the day, my odor was quite powerful," she comments (*ABM* 70). As discussed in the previous chapter's section on odor, I argue that in Kincaid's works body smells represent transgressions of the exterior borders of the human body. Through odor, bodies make themselves known and may subvert the order of established power structures. Here, Xuela's taking pleasure in her own vulva's odor, as she removes it from parts of her body that are out of sight (still under her clothing) and transfers it into the open, transports parts of her inside to the outside. Xuela thus emphasizes her genitals and affects the space around her. The adjective "powerful" initially describes only Xuela's odor, but metonymically refers to Xuela and her body as a whole. In this way, Xuela dominates the space around her. The male gaze of Monsieur LaBatte may participate here, but it has no power over Xuela as she counters it with her own in combination with her odor.

---

232 My analysis of Xuela's precarity as integrating her with historical vulnerabilities reads the "inevitable" in this vein; see the previous section on "Abortive Practices."

In the subsequent scene, *The Autobiography of My Mother* presents the approximation of Xuela and Jacques LaBatte through a symbolism of light and dark. The encounter begins with a setting at dusk: "The sun had not yet set completely; it was just at that moment when the creatures of the day are quiet but the creatures of the night have not quite found their voice" (*ABM* 69). Xuela identifies this in-between with a sense of loss – of mother, home, and love, i. e., the loss of a sense of belonging. The text thus locates the ensuing events in an interstitial space. Decoding the symbolism of light here, Xuela would be associated with the interstice between light and dark, whereas Monsieur LaBatte is a "creature of the night." He remains quiet and at bay while the light fades and night falls. Only then he approaches and uses his voice to ask Xuela to fully remove her clothes (*ABM* 69–70). Yet again, the girl does not yield to his request. Encompassed by the darkness associated with Monsieur LaBatte,[233] Xuela does not give up the space she had claimed by exuding her own body odors. Instead, she rebuffs him, as "it was too dark, I could not see" (*ABM* 70). Monsieur LaBatte then takes the girl to his private room, which is generally dark, only lit by a small lamp. Reading this interplay of light and darkness in this scene, Monsieur LaBatte's room might be a space determined by him, yet with the dusky light the lamp provides, the room is affected by Xuela's association with the in-between of twilight. In this, the symbolism of illumination contributes to the representation of Xuela as not without presence – and by extension agency – in this setting, which determines my reading of the following sex scene.

Xuela is now thrilled by the anticipation of what is to come. Again, the choice of words here hardly evaluates the situation: Xuela describes herself as "enthralled" (*ABM* 71), which points to her interest in the situation, possibly a desire to have a curiosity satisfied. When Xuela then begins to relate the actual penetration of her body by Jacques LaBatte, she uses the very terms she previously chose to introduce the whole episode with earlier:

> And the force of him inside me, inevitable as it was, again came as a shock, a long sharp line of pain that then washed over me with the broadness of a wave, a long sharp line of pleasure: and to each piercing he made inside of me, I made a cry that was the same cry, a cry of sadness, for without making of it something it really was not I was not the same person I had been before. He was not a man of love, I did not need him to be. When he was through with me and I with him, he lay on top of me, breathing indifferently, his mind was on other things. (*ABM* 71)

---

233 In the ensuing sexual relations, too, Monsieur LaBatte is associated with the nights Xuela spends with him, while the girl's time during the day is allocated to Madame LaBatte (*ABM* 74).

Although the male role here is portrayed as forceful and as physically breaching, Xuela's experience is not defined by negativity. Pain is immediately coupled with pleasure. Xuela emphasizes the physical enjoyment she derives from sexual intercourse, when she points out that a male body can make her feel "a wholeness with a current running through it, a current of pure pleasure" (*ABM* 70–71). The sensations of pain and pleasure are linked here by their immediate proximity, as they occur in the same sentence; by their unity in their definitions as "a long sharp line of"; and not least by the alliterative connection of the two words. Xuela's physical sensations thus are doubled in this seemingly contradictory pair of feelings.

The cries she utters, however, are not moans of arousal but of grief for the loss of her former self. Her audible expression of it once more speaks to her active participation in the scene. Sexual intercourse here marks a before and after in Xuela's life. This again is brought forth matter-of-factly, rather than in evaluative fashion. Xuela closes this brief account of her first sexual encounter by highlighting an equality in the couple: Monsieur LaBatte is not a lover and he is indifferent, but Xuela herself is indifferent to this circumstance, as she is focused on the physical here. Moreover, they are "through" with each other, i. e., Xuela emphasizes her participation in the sexual act, rather than denoting possible victimhood in her body having solely been used by another for their pleasure.

The physical violence of the encounter is related retrospectively, when Xuela is back in her room after the fact by herself. She here reflects on the lasting "ache" she feels in several body parts:

> The place between my legs ached, my wrists ached; when he had not wanted me to touch him, he had placed his own large hands over my wrists and kept them pinned to the floor; when my cries had distracted him, he had clamped my lips shut with his mouth. It was through all the parts of my body that ached that I relived the deep pleasure I had just experienced. (*ABM* 72)

These details evoke a S/M dynamic with Monsieur LaBatte in the dominant position. If not for the repeated connection of ache and pleasure, this scene might indeed indicate Jacques LaBatte forcing sexual relations, i. e., raping Xuela or at least coercing sexual favors from her. Nonetheless, while the details of violence here suggest this reading, the passage does not victimize Xuela. The "ache" she feels points to a past pleasure itself in that it ambivalently denotes the physical pain and a yearning for the return of this touch.

In her analysis of the sexual scene in question, Rhonda Cobham understands the absence of victimhood as a rhetoric strategy on the part of autodiegetic narrator Xuela. She argues that "Xuela's self-directed monologue utilizes a language of negation that gives her a perverse strength in relation to all forces that seek to dominate her through language or action. […] But the language Xuela uses to describe their [Xuela and Jacques LaBatte's] encounter manipulates negatives in ways that create the illusion of agency for the raped school girl" (876).[234] Thus, what I read as agency in Xuela's sexual involvement, Cobham evaluates as belated rhetoric that refuses to admit to the girl's defeat. However, while Cobham's observations on Xuela's narrative strategies are compelling, I add that a consideration of space – as shown – supports a positioning of Xuela beyond the role of the victim.

Focusing on Xuela's resistances and participation then allows us to read the strength with which Xuela faces her vulnerabilities. On the conjunction of gaze and power, bell hooks holds that looking back is a means of resistance to domination – as is the case for Xuela in the LaBattes' yard that night, I argue. Taking Xuela's narrative and her sexual agency seriously does not exculpate the perpetrator Jacques LaBatte, but it may free Xuela from our readerly "gaze" framing her as the powerless victim. "We do more than resist. We create alternative texts that are not solely reactions. As critical spectators, black women participate in a broad range of looking relations, contest, resist, revision, interrogate, and invent on multiple levels," hooks reminds us ("Oppositional Gaze" 128). I suggest to read Xuela's looking back at Jacques LaBatte as well as her later narrative of the ensuing events within the variable parameters of resistance as detailed by hooks in this quote. Reading the sex scene in a context of S/M thus means reading agency in Xuela's seeming passivity. As Anne McClintock writes, "[t]o argue that in consensual S/M the 'dominant' has power, and the slave has not, is to read theater for reality" ("Maid to Order" 87).

Finally, Cobham concludes that "by the end of the encounter it is no longer clear who has seduced whom; who has used whom to assuage whose passion" (878), with which my analysis here agrees. Ultimately, it is this confusion and ambiguity that makes a reading of Xuela's sexual agency possible. In any case then, whether we read the scene as rape or S/M, this sexual encounter in *The Autobiography of My Mother* evokes such differing interpretations because it is mediated by literary text. Sabine Sielke, in her study of the rhetoric of sexual violence in literature, regards rape as a "figure of power relations" (3). Sielke emphasizes that texts relate to rape in highly mediated ways and that "[t]he determinate given raw material of rape narratives […] is not necessarily rape. […] any historical issue is inseparable from its various cultural representations" (3). In the case of *The Autobiography of My Mother*, the indeterminacy of sexual violence highlights its status as a representation. Reading it as an enactment of predetermined ("inevi-

---

234 Carine Mardorossian agrees with Cobham and adds that the illusion of Xuela's sexual agency in the "seduction scene dismantles our assumption about both rape and sex because it turns the young girl into both victim and instrument of sexual subordination" ("Rape by Proxy" 124).

table") roles makes Xuela and Jacques LaBatte's sexual relation a presentation of rehearsed roles. Xuela notes this when she observes that "[h]e was behaving in a way that he knew well" (*ABM* 77). Monsieur LaBatte thus assumes the paradigmatic role of the exploiter.

This staging of violence (whether we explicitly read the scene as S/M or not) highlights the historical dimension of systemic sexual violence and provides space for a reclamation of Xuela's self. Considering the history of rape in slavery, this is no less than a reclamation of the enslaved female body, which was – among other aspects – victimized via sexual violation. Angela Davis describes the sexual assault of Black women during slavery as "reducing her to the level of her *biological* being" (96, emphasis in original), i. e., rape constituted an assault on the subjectivity as human. Davis regards this as one aspect of the "terroristic texture of plantation life" (96). Taking this into account, I read the representation of sexual violence in *The Autobiography of My Mother* really as a negotiation of historical oppression of subjectivity during slavery. Carine Mardorossian comes to a similar conclusion:

> By performing the role scripted for her by her oppressor's colonialist fantasies, she [Xuela] exposes the ideological molds that contain his world as strings that control him as much as they subjugate her. Far from "naturalizing" the sexualized encounter, her apparent complicity exposes the rape as a scene in a play of power whose arbitrary rules she can perform again and again and from which she thus learns to distance herself. ("Rape by Proxy" 126)

This examination of personal and collective past through repetition and reenactment also is at stake when towards the end of the book Xuela enters a relationship with Philip Bailey, an English doctor.

The Autobiography of My Mother introduces Philip by reference to the typical interests of a representative of the British empire: he is infatuated with cultivating flowers and plants, landscape, ruins, and books (*ABM* 143–144, 146), i. e., he is fascinated with manipulating the original resources of the Caribbean, with naming and recording them in written form, as well as with the past. His hobbies recall the plant hunters of the late nineteenth and early twentieth century, who claimed and colonized by naming plants.[235] In their desire to "order" and transplant botanical species for profit, these acts of colonization from afar are structurally the same as those of the transatlantic slave trade. *The Autobiography of My Mother* formulates Philip's fascination with plants and lands quite explicitly as an inheritance of colonization:

235 See chapter 3.2 on further discussions of plant hunters and colonization in Kincaid's *My Garden (Book):*. See also De Ferrari, who reads Philip as "a sort of eighteenth-century botanist" (155).

> He had an obsessive interest in rearranging the landscape: not gardening in the way of necessity, the growing of food, but gardening in the way of luxury, the growing of flowering plants for no other reason than the pleasure of it and making these plants do exactly what he wanted them to do; and it made great sense that he would be drawn to this activity, for it is an act of conquest, benign though it may be. (*ABM* 143)

Philip is thus characterized by an inert desire to dominate and mold his surroundings to his liking. Yet in the early twentieth century he is but a remnant of the British colonizers who transplanted and exploited human beings during slavery and in Caribbean plantation societies. In this sense, Philip is rather a reminder of the past, a blunted "benign" version, yet still a representation of the desires of British colonialism. This status of the remnant is also indicated by his interest in "ruins, not the kind that are remains of a lost civilization, but purposely built decay" (*ABM* 143). Philip is interested not in the actual past, but in a manipulated version of it. This manipulation turns the past into a commodity, not only through the ruins as purported artifacts, but also through the books that picture them and make them fit for consumption from a distance.

Xuela's relationship with the colonizer Philip is an extended repetition of her relations with Jacques LaBatte in her earlier years, yet with the difference that the stakes are much higher: Philip is a white Englishman whereas Jacques LaBatte was an upper-class Afro-Caribbean man; Xuela and Philip get married, while she only had a clandestine sexual relationship with Monsieur LaBatte; the sexual intercourse with Philip appears to be even more violent than with Monsieur LaBatte; and finally Xuela much more overtly orchestrates the relationship and thus glories in her own sexual agency, which in these contexts metonymically represents a small triumph of colonized over the colonizer.

The parallels between the two relationships begin with the circumstances that lead Xuela to meet the men in the first place. What brings Xuela to be with Philip, a doctor living in Roseau, is an arrangement that Xuela's father made with a friend of his (*ABM* 128). Xuela is to be employed as a domestic worker, living in a room set apart from the main house, originally built to accommodate a nurse to the doctor's office (*ABM* 150–151) – similar to the housing arrangement on the LaBattes' property. Philip also is married and his relationship with Xuela would initially count as an affair with "the help." Eventually, however, after the death of this wife, Xuela and Philip get married and Xuela lives in the main house.

The parallels of Xuela's affair with Monsieur LaBatte and Philip do not stop here. The relationship with Philip is also narrated through the spaces it is set in. The first scene that establishes sexual desire between Xuela and Philip takes place in the doctor's office. Philip examines his employee for a persistent cough, in the course of which she removes her garments, including the muslin strips she uses to bandage her breasts to ease an irritation of her nipples. This room, in which Xuela's body is bared to view, is thoroughly defined by Philip's presence. His workspace is filled with paraphernalia that represent him, such as "white enamel basins which held syringes and needles and forceps" (*ABM* 148). Moreover, this scene again is accompanied with distinct detailing of the lighting in the room, which reflects the relationship of the characters within that space:

> [T]he room had windows on three sides, the windows had adjustable wooden slats; the wooden slats were tilted half open and the sunlight came in through them, measured, each shaft three inches wide, and some of them fell diagonally across another part of the floor and then bent up against the wall and ended halfway there, and it gave the room a strange atmosphere, the pattern of shading and light, a fully clothed man, a woman explaining why she bandaged her breasts (*ABM* 148).

The light and the window slats fragment the room, while the interlacing pattern of light and dark separates and unites. Such patterning is repeated in the opposition of the clothed man and the half-naked woman, her bandages interlaced on her skin. Moreover, the interlacing and bending against the wall might be read as foreshadowing the later explicit sex scene in which the bodies of Xuela and Philip do the same. For now, although in his very own space, Philip turns away from it and towards the sunlight outside the room, as "all of a sudden he must have felt excited for he walked away from me [Xuela] and looked through one of the half-closed shutters" (*ABM* 148). Turning towards daylight, Philip turns away from Xuela's body, which he will nonetheless visit again in more dimmed lighting.

While this first scene of physical closeness between Philip and Xuela shows significant parallels to that with Monsieur LaBatte as discussed earlier, it differs in the details that ultimately position Xuela as much more self-determined than with Jacques LaBatte. For instance, this later passage leaves it open who initially approaches whom. On the one hand Xuela very innocently seeks out her employer's expertise on a health issue. The examination of her body then leads to lust on the doctor's part, which might be evaluated as unprofessional and inappropriate behavior. When the examination of her chest, however, leads to a focus on her breasts, which apparently suffer from a medical issue, Xuela herself introduces sexuality to the scene when she explains: "My breasts then were in a constant state of sensation [...] [that] ceased only when a mouth, a man's mouth, was clamped tightly over them and sucking" (*ABM* 147). Emphasizing not the ailment, but sexual stimulation as the cure, Xuela herself turns the attention to sexual aspects of her body. She also explicitly notes the absence of ulterior motives on her part when she mentions that she moved the muslin straps "as if I was alone, and this was because I was in the presence of a doctor, not because I wanted him to find it interesting in any way" (*ABM* 147); but she simultaneously tells Philip about erotic acts as easing her condition. Xuela mixes up the professional and the erotic, so that this scene might just as well be read as seductive, i. e., as inducing Philip's unprofessional desires. Ultimately, both readings are viable, and the text leaves it to interpretation who incites the following sexual relations.

The concrete sexual acts then do not take place in his rooms but in hers. Here, ostensibly the scene from the LaBattes' yard repeats. Night has already fallen outside when Philip enters Xuela's room, apparently believing that he is crossing the distance between her and his living space in order to show her a book (*ABM* 151).[236] He finds Xuela touching herself intimately. Xuela strokes her pubic hair and does not stop when Philip comes upon her. Instead, when he calls her name, she begins to remove her nightgown under which she is naked; but it gets trapped on the mass of her pinned-up braids, so that Xuela's body is fully exposed to Philip's gaze, while her face is hidden and her hands and arms are tied up in the garment above her head. This position already anticipates what is to come, when Xuela and Philip later enact the very same position. Philip begins to kiss Xuela and helps her out of her gown, which leads to the following sexual scene:

> I turned around and removed his belt, and using my mouth I secured it tightly around my wrists and I raised my hands in the air, and with my face turned sideways I placed my chest against the wall. I made him stand behind me, I made him lie on top of me, my face beneath his; I made him lie on top of me, my back beneath his chest; I made him lie in back of me and place his hand in my mouth and I bit his hand in a moment of confusion, a moment when I could not tell if I was in agony or pleasure; I made him kiss my entire body, starting with my feet and ending with the top of my head. The darkness outside the room pressed against is four sides; inside, the room grew smaller and smaller as it filled up almost to bursting with hisses, gasps, moans, sighs, tears, bursts of laughter. (*ABM* 154–155)

Xuela's being tied up with a belt indicates a bondage scene. In this sexual staging, she assumes the role of the submissive, while Philip acts out the dominant part. In an S/M binary, Xuela is the slave, while Philip is the master. Still, by Xuela's repetition of "I made him," it is clear that she is the choreographer of this act, which upturns the ostensible binary of dominant and submissive. This fluidity and negotiation of power through sexual performance exemplifies what McClintock calls "the scandal of S/M," namely its "provocative confession that the edicts of power are reversible" ("Maid to Order" 87).

Xuela finds her pretend-submission delicious, as she realizes when she first is trapped in her nightgown. "I became eternally fascinated with how I felt then," Xuela notes. "I felt a sensation between my legs [...] the feeling was a sweet, hollow

236 Xuela, however, refuses to look at the book, an object identified with Philip the colonizer. She simply dismisses this advance. See *ABM* 151.

feeling, an empty space with a yearning to be filled" (*ABM* 153–154).[237] Gary and Kimberley Holcomb, who also read the sex scenes with Philip in terms of S/M practices, similarly note Xuela's agency within her supposed submission: "Xuela plays with received roles in the above scene, subverting the relation of passivity and bondage to activity and agency or dominance: she is the one 'fascinated' by the feelings she takes from exposure and restraint; she is actively aroused and in pursuit of pleasure rather than object of someone else's desire" (13). Xuela overpowers Philip, as is also apparent in his "drowning" in her presence when he speaks to her "as if he were imprisoned in the sound of my [Xuela's] name" (*ABM* 153). Xuela's overwhelming presence is also once again signified by the room's illumination. Her private space is dimly lit by a lamp that links Xuela to her maternal genealogy when she notes, "it had been my mother's lamp and would have been the last lamplight she saw, because it was the lamp that lit the room at the time she died, which was the time I was born; and by this lamplight, too, she would have seen my father's face as he lay on top of her, just before he withdrew himself from inside her" (*ABM* 150–151). The lamp thus connects Xuela with her personal history in that it is the first light Xuela would have seen at birth. Moreover, having sex in the light of this lamp performatively links Xuela to her lost mother who did the same. Xuela thus stands in the light of her inheritance, i.e., the room she and Philip now occupy is defined by Xuela's past and female ancestral lines that the lamp signifies.

As the instigator and orchestrator of the S/M performance, Xuela forces Philip into the role of the colonizer, enacting his forceful perpetration on her body. Through the stylized form of S/M, making use of one of its typical props – the bonds – this scene lays bare the paradigmatic distribution of roles here. And as the lamp indicating female ancestry also suggests, Xuela's binding presents her as a stand-in for the colonized female body. Guillermina De Ferrari logically contends that Xuela's "body is never a fully private entity, but is instead a site where cultural scripts originally inherited from colonial history are constantly redeployed" (147). The very reenactment of sexual violence through S/M moreover brings about its subversion in that "S/M performs social power as scripted, and hence as permanently subject to change" (McClintock, "Maid to Order" 89).

Xuela's dominating the performance relocates Philip to the role of the subjugated. Xuela generally refuses to regard her husband in his personal particularities; she solely sees him in his role as representing "the colonizer" (see *ABM*

---

237 See also De Ferrari who reads the nightgown's concealing of Xuela's face as doubling her role into "masked 'master' and tied 'slave'" (160).

219).[238] She thus defines their lives together through her negotiation of colonial history, which they perform in their sexual and everyday interactions. Although not an enslaver or conqueror himself, being regarded as such is his legacy. "[B]y the time Philip was born, all the bad deeds had been committed; he was an heir, generations of people had died and left him something" (*ABM* 146). Philip is an heir to the "victors" (*ABM* 217); his inheritance is social and economic power. But – as Xuela makes clear – it also the shameful association with the perpetrations of colonialism. "He was an heir, and like all such people the origin of his inheritance was a burden to him," Xuela concludes (*ABM* 220). Philip's love and devotion to Xuela might be an attempt to relieve himself of this burden, but his wife refuses him such ease when she keeps him in the role of the colonizer.

The S/M scene is thus integrated into a framework of Philip's subjugation by Xuela. She moreover isolates him from society by moving him to the mountains (*ABM* 221), as she does from herself by withholding her presence from him (*ABM* 219). She lets him act as a colonial naturalist and explorer, but manipulates his efforts through purposeful sabotage. When he keeps small animals in terraria to observe them in different environments, Xuela obstructs Philip's small-scale colonial efforts by "releas[ing] whatever individual he held in captivity, replacing it with its like, its kind: one lizard replaced with another lizard [...]. He was so sure inside himself that all the things he knew were correct" (*ABM* 222). Xuela thus keeps Philip in his role of the colonizer, yet rendered unsuccessful by her subversions, his very epistemological methods are rendered futile and through her manipulations they are rather turned into child's play than the knowledge production he believes himself to accomplish.

In conclusion, S/M practices in Kincaid's works as discussed in this section enable the texts' protagonists to negotiate power structures that affect all aspects of their lives. They find symbolic expression in scenes of physical pleasure and pain – be it a questioning of gender roles as dictated by all-powerful mother figures and their prescriptions of (sexual) morality in *Annie John* or *Lucy*, or the power imbalances that are violently inscribed on the collective female body of color during colonialism as in *The Autobiography of My Mother*. "S/M is haunted by memory. By reenacting loss of control in a staged situation of excessive control, the S/Mer gains symbolic power over perilous memory. By reinventing the memory of trauma, S/M affords a delirious triumph over the past, and from this triumph an orgasmic excess of pleasure" (McClintock, *Imperial Leather* 100). As shown, in the case of

238 De Ferrari compellingly reads this in another sexual scene, when Xuela refuses to listen to Philip's childhood stories and instead orders him to perform oral sex on her. "Xuela refuses him the right to a personal narrative by turning him into a sexual slave. The marks left on his face by her pubic hair evoke the branding of slaves as property" (159).

Xuela in *The Autobiography of My Mother*, the final (and most overt) S/M scene repeats sexual violence in a staging that inverts original power relations. The repetition of violent sexual acts from the scene with Jacques LaBatte to Philip Bailey thus points to a continuity of historical violation of female bodies in personal contemporary experiences.

Still, the two instances of violent sex in *The Autobiography of My Mother* significantly differ in that Xuela's sexual encounter may reasonably be examined as to whether it is consensual or forced, whereas in the later sex scene with Philip, Xuela is the director of the event. In this, the protagonist's sexual agency is developed from questionable to obviously manifested and acted out. This difference reverses both male and colonial sexual power over her (collective) body. Thus gaining symbolic power over the past and the historical violated female body, Xuela's actions also question the fixity of colonial binaries in general.[239] Reenactments and bodily performance thereby stage historical drama and decenter the previously ruling power by placing Xuela's sexual agency and pleasure at its heart. In this, I conclude, the S/M scene between Xuela and Philip destabilizes the scripts of colonial power through bodily performance.

---

239 Gary and Kimberly Holcomb also read diffusion of binaries in Xuela's S/M orchestration: "Ultimately, S/M permits Xuela to rewrite the history of cruelty and barbarism enacted on her heritage, the centuries of a pain legally regulated, narratologically scripted, and epistemologically determined by the master by possessing, directing, controlling and exploring the uncharted island of her own pain. In shifting agency from the master's to the slave's body, Xuela simultaneously undermines the master's identity, reducing him to a near ghost, and changes the meaning of pain itself, from its association with colonial punishment and lust to a means of pleasuring the slave's desiring body; the distinction between slave and master blurs" (17).

Epilogue: Permanent Impermanence

In a recent essay, Jamaica Kincaid allegorizes the political situation in the United States of 2020 with the image of a snow dome. The short text "Inside the American Snow Dome" was written on the occasion of Joe Biden's election to be the forty-sixth president of the United States.[240] The essay celebrates the political change of the election of a Democratic president and derides the nationalist claim to permanence which links whiteness with Americanness in a presumed historical disposition. Using the image of the snow dome, the text applies Kincaid's general interest in dissipating fixity and totality to the current situation in which it was written. In this, it well exemplifies the results of my observations on how Kincaid's works understand representations of both the past and present: nothing is permanent, and the very proposal of permanence contains the promise of impermanence. I have shown this with my analyses of how Kincaid's use of media engages with the fixity of colonial representations of history. In formally revising the media they invoke, Kincaid's texts and images signify hitherto hidden perspectives on the past.

In "Inside the American Snow Dome," Kincaid presents Donald Trump's period in office as an extension of colonial history, whose beginning she traces to when Christopher Columbus set sail on his voyages of "discovery" in 1492. She describes Trump's presidency as fixing her within "a snow dome that resembles the United States of America." Yet, while the snow dome appears to freeze in permanence, she argues, impermanence is inherent to its ontology:

> [A] snow dome is an object, dome-like in shape, resting on a flat piece of material that is fitted to it and sealed perfectly to its base. The entire structure is made of a material that is easily shattered. [...] Scenes of one kind or another are created and fixed to the bottom of the dome. Flakes of something white made to resemble snow are settled at the bottom of the dome, and when the dome is shaken, as it often is by a playful hand passing by, the flakes rush up in a flurry around the scene that has been fixed to the bottom of the dome. All the figures and objects are lost in a blur of the pretend snow, they are consumed by it, and for a moment, it seems as if this will be the new forever: they will never be seen clearly again. Then the false snow slowly settles back to the bottom of the dome and everything returns to the way it was. The scene remains just as it was before, fixed, fixed, and fixed!

Although fixity is implied in the figurines glued to the bottom, and the glass dome implies totality by encasing it all, Kincaid's description of the snow dome simulta-

240 First published in the Swedish newspaper *Dagens Nyheter*, republished online in *The Paris Review* on November 11, 2020. I cite the essay from the online version, which has no page numbers.

https://doi.org/10.1515/9783111027500-007

neously points to the inherent impermanence of what it contains. The fake snowflakes dissipate any fixity in swirling around the glass. Moreover, applying a change of perspective and looking at the snow dome from the outside reveals the fragility of the whole object, since the glass shatters easily. As such, both fixity and totality are represented as carrying their own destruction within themselves.

I read this snow dome not only as an allegory of Trump's America, but also as a representation of Kincaid's conception of representations in general: reading the snow dome as a sign or a received representation often means emphasizing one of its conditions, to wit, the moment when it appears to us in its fixed state. Kincaid's work on the sign shakes up this fixity and presents snowflakes both as already present and settled within the dome, and as meant to be shaken up in the praxeology of the object. In their immobile state, the snowflakes might be thought in analogy to Henry Louis Gates's repressed meanings of a word that lie in wait to be activated by Signifyin(g). Once shaken up, the signs might settle again, as the flakes in the snow dome do, but the fixity the essay proclaims in three exclamations does not ensue. Indeed, although momentary stillness gives the impression of fixity, the swirl of flakes always settles in new and different arrangements. In this sense, the chaos here – as with Édouard Glissant's poetics of relation – has a productive and renewable quality.

In this conception, the possibility of shaking up form and thus significations – be it of colonial history, current politics, or gendered or racialized bodies, etc. – is inherent to any representation. Finally, Kincaid points to the fragility of the representation, which also means it may cease to serve its purpose: the snow dome may shatter. And at the end of her essay, it indeed falls and shatters. Kincaid then describes herself as picking up only the pieces of the figurines she favors. She mends them, but grounds the rest to dust with a hammer. This destructive image contains the positive outlook that a representation can be remade. Its seeming fixity and totality do not have to be accepted, but can be taken apart and made anew in useful ways, as Kincaid's poetics of impermanence demonstrate throughout her works.

In this, the allegory of the snow dome read as a representation of Donald Trump's presidency really is a comment on power structures whose assumed permanences (since colonial times, as the reference to 1492 implies) have always been fraught. Allegorically shattering the snow dome, Kincaid simultaneously decenters the assumed superiority of whiteness when her essay points out that definitions of whiteness, too, were not permanent throughout history, such as the racialization of Catholics in the Elizabethan era. In this way, the essay shows that whiteness – as well as Americanness – is just as prone to external definition (here through racialization) as is any other social or ethnic group inside and outside the United States. Seen in this light, through a history of its external definitions, whiteness proves to be as fragile as a snow dome's glass, and any claim to white exceptionalism shatters just like the dome itself. Kincaid's essay shows that American history is thus not a history of whiteness, but really a history of immigration (of multitudes of different people), of segregation (between various kinds of groups), and of being "not free." In deconstructing the key assumption of American nationalism, the essay in fact shifts the focus to African Americans, who bring an intimate understanding to this different view of American history. Hence, Kincaid places them at the center of Americanness – "we are the true Americans" – emphasizing the presence, importance, and value of Black people in the United States, which white nationalism and racism continually seek to repress.

In the short text "Inside the American Snow Dome," Jamaica Kincaid applies the same fundamental thought of impermanence to the snow dome and to the recent history of the United States. What is at stake here is contemporary white racism against Black people, which Kincaid identifies as the continuation of colonial history. Her text negates the assumption of permanence, as do her poetics throughout her oeuvre. It is in this "permanent impermanence," as she puts it, that the potential lies for the emergence of other voices and perspectives that the totality of assumed Western superiority – both its history and its present – has always sought to silence. In revealing white nationalism as an extension of colonialist ideology, Kincaid's strategies of destabilizing structures of fixity in her engagements with colonial history prove equally useful as responses to the totalizing forces of the present. They offer strategies for recognizing the multiplicity of peoples and histories that make up the United States, in the same way that the offer strategies for resisting History with a capital *H* in postcolonial Caribbean contexts.

Works Cited

Abel, Elizabeth. "Domestic Borders, Cultural Boundaries: Black Feminists Re-View the Family." *The Familial Gaze*, edited by Marianne Hirsch, UP of New England, 1999, pp. 124–152.

Adams, Michelene. "Jamaica Kincaid's *The Autobiography of My Mother:* Allegory and Self-Writing as Counter Discourse." *Anthurium: A Caribbean Studies Journal*, vol. 4, no. 1, 2006, n. pag.

Alexander, M. Jacqui. *Pedagogies of Crossing: Meditations on Feminism, Sexual Politics, Memory, and the Sacred.* Duke UP, 2005.

Alexander, Simone. "I Am Me, I Am You: The Intricate Mother-Daughter Dyadic Relationship." *Mother Imagery in the Novels of Afro-Caribbean Women*, U of Missouri P, 2001, pp. 45–95.

Alexander, Simone. "Postcolonial Hauntings: Ghostly Presence in Jamaica Kincaid's *The Autobiography of My Mother.*" *Wagadu: A Journal of Transnational Women's and Gender Studies*, vol. 19, Summer 2018, pp. 107–130.

Anatol, Giselle Liza. "Speaking in (M)Other Tongues: The Role of Language in Jamaica Kincaid's *The Autobiography of My Mother.*" *Callaloo*, vol. 25, no. 3, 2002, pp. 938–953.

Anker, Elizabeth S. "'Commonly Human': Embodied Self-Possession and Human Rights in Jamaica Kincaid's *The Autobiography of My Mother.*" *The Routledge Companion to Literature and Human Rights*, edited by Sophia McClennen and Alexandra Schultheis Moore, Routledge, 2015, pp. 37–45.

Ankersmit, F. R. *Historical Representation.* Stanford UP, 2001.

Apollodorus. *The Library. Volume I: Books 1–3.9.* Translated by James George Frazer, Harvard UP, 1995.

Arbino, Daniel. "Loss of Mother, Loss of Self: Orphanhood in Jamaica Kincaid's *The Autobiography of My Mother.*" *Reading/Speaking/Writing the Mother Text: Essays on Caribbean Women's Writing*, edited by Cristina Herrera and Paula Sanmartín, Demeter Press, 2015, pp. 215–234.

Astor, Maggie. "Dove Drops an Ad Accused of Racism." *The New York Times*, 8 Oct. 2017, www.nytimes.com/2017/10/08/business/dove-ad-racist.html. Accessed 28 Mar. 2019.

Ayre, Michael. *The Caribbean in Sepia: A History in Photographs 1840–1900.* Ian Randle Publishers, 2012.

Azima, Rachel. "'Not-the-Native': Self-Transplantation, Ecocriticism, and Postcolonialism in Jamaica Kincaid's *My Garden (Book).*" *Postcolonial Studies and Ecocriticism*, edited by Anthony Vital and Hans-Georg Erney, special issue of *Journal of Commonwealth and Postcolonial Studies*, vol. 13, no. 2, 2006 and vol. 14, no. 1, 2007, pp. 101–119.

Babilon, Daniela. *The Power of Smell in American Literature: Odor, Affect, and Social Inequality.* Peter Lang, 2017.

Bakhtin, Mikhail. *The Dialogic Imagination. Four Essays.* Translated by Caryl Emmerson and Michael Holquist, U of Texas P, 2008.

Bakhtin, Mikhail. *Rabelais and His World.* Translated by Helene Iswolsky. M.I.T. Press, 1968.

Ball, John Clement. "Drickie Potter and the Annihilating Sea: Reading Jamaica Kincaid's Waves of Nothingness." *Literature for Our Times: Postcolonial Studies in the Twenty-First Century*, edited by Bill Ashcroft et al., Rodopi, 2012, pp. 203–220.

Banerjee, Chinmoy. "The Erotics of Resistance and Postcolonial Self-Fashioning in *Annie John.*" *ARIEL: A Review of International English Literature*, vol. 33, no. 2, 2002, pp. 33–56.

Baraka, Amiri. *The Autobiography of LeRoi Jones.* Lawrence Hill Books, 1997.

Barthes, Roland. *Camera Lucida: Reflections on Photography.* Translated by Richard Howard, Vintage Books, 2000.

https://doi.org/10.1515/9783111027500-008

Barthes, Roland. "The Death of the Author." *Image, Music, Text.* Translated by Stephen Heath, Hill and Wang, 1977, pp. 142 – 148.

Barthes, Roland. *The Pleasure of the Text.* Translated by Richard Miller, Hill and Wang, 1975.

Barthes, Roland. "The Rhetoric of the Image." *Image, Music, Text.* Translated by Stephen Heath, Hill and Wang, 1977, pp. 31 – 51.

Barthes, Roland. "Theory of the Text." *Untying the Text: A Post-Structuralist Reader*, edited by Robert Young, Routledge, 1990, pp. 31 – 47.

Baucom, Ian. *Specters of the Atlantic: Finance Capital, Slavery, and the Philosophy of History.* Duke UP, 2005.

Baugh, Eddie. "Literary Theory And The Caribbean: Theory, Belief And Desire, or Designing Theory." *Journal of West Indian Literature*, vol. 15, no. 1 – 2, 2006, pp. 3 – 14.

Baym, Nina, and Robert S. Levine, editors. *The Norton Anthology of American Literature*, shorter 8th ed., Norton, 2013.

Bazin, André. "The Ontology of the Photographic Image." *What Is Cinema?* Edited and translated by Hugh Gray, U of California P, 1967, pp. 9 – 16.

"Benighted." *Merriam-Webster's Dictionary*, Merriam-Webster, www.merriam-webster.com/dictionary/benighted. Accessed 25 Oct. 2017.

Benjamin, Walter. *The Arcades Project.* Translated by Howard Eiland and Kevin McLaughlin, Belknap P of Harvard U, 1999.

Benjamin, Walter. "Little History of Photography." *Selected Writings: 1927 – 1934*, edited by Michael W. Jennings et al., vol. 2, Belknap P of Harvard U, 2001, pp. 507 – 530.

Bennett, Louise. *Anancy and Miss Lou.* Sangster's Book Stores, 1979.

Bhabha, Homi K. *The Location of Culture.* 1994. Routledge, 2004.

Bickford, Ian. "'Dead Might Not Be Dead': Milton in the Americas and Jamaica Kincaid's Flat World." *Modern Philology: Critical and Historical Studies in Literature, Medieval Through Contemporary*, vol. 11, no. 4, 2014, pp. 862 – 878.

Birbalsingh, Frank. "Jamaica Kincaid: From Antigua to America." *Frontiers of Caribbean Literature in English*, edited by Frank Birbalsingh, St. Martin's, 1996, pp. 138 – 151.

Bloch, Ernst. *Erbschaft dieser Zeit.* 1935. Suhrkamp Verlag, 1973.

Bloch, Ernst. "Nonsynchronism and the Obligation to Its Dialectics." Translated by Mark Ritter, *New German Critique*, vol. 11, Spring 1977, pp. 22 – 38.

Bloom, Harold, editor. *Jamaica Kincaid.* Chelsea House, 2008.

Bolaki, Stella. "'Mourning Remains': The Poetics and the Politics of Loss in Jamaica Kincaid's *At the Bottom of the River* and *Lucy*." *Come Weep with Me: Loss and Mourning in the Writings of Caribbean Women Writers*, edited by Joyce C. Harte, Cambridge Scholars, 2007, pp. 164 – 180.

Bonetti, Kay. "An Interview with Jamaica Kincaid." *The Missouri Review*, vol. 15, no. 2, 1992, pp. 124 – 142.

Bourdieu, Pierre. "The Social Definition of Photography." *Photography: A Middle-Brow Art.* Translated by Shaun Whiteside, Polity Press, 1996, pp. 73 – 98.

Bouson, J. Brooks. *Jamaica Kincaid: Writing Memory, Writing Back to the Mother.* State U of New York P, 2005.

Boxer, David, and Edward Lucie-Smith. *Jamaica in Black and White: Photography in Jamaica c.1845 – c.1920: The David Boxer Collection.* Macmillan Caribbean, 2013.

Boyagoda, Randy. "Five in the Colonies: Enid Blyton's Sri Lankan Adventures." *The Paris Review*, 25 June 2012, www.theparisreview.org/blog/2012/06/25/five-in-the-colonies-enid-blyton%E2%80%99s-sri-lankan-adventures/. Accessed 29 July 2017.

Braithwaite, Alisa K. "Reimagining Migration through the Act of Writing in Anglophone Caribbean Women's Narrative." *Tulsa Studies in Women's Literature*, vol. 32, no. 1, 2013, pp. 143–158.
Brancato, Sabrina. *Mother and Motherland in Jamaica Kincaid.* Peter Lang, 2005.
Brathwaite, Edward Kamau. *History of the Voice: The Development of Nation Language in Anglophone Caribbean Poetry.* New Beacon Books, 1984.
Braziel, Jana Evans. "Alterbiographic Transmutations of Genre in Jamaica Kincaid's 'Biography of a Dress' and *Autobiography of My Mother.*" *A/B: Auto/Biography Studies*, vol. 18, no. 1, 2003, pp. 85–104.
Braziel, Jana Evans. "'Another Line Was Born…': Genesis, Genealogy, and Genre in Jamaica Kincaid's *Mr. Potter.*" *Jamaica Kincaid and Caribbean Double Crossings*, edited by Linda Lang-Peralta, U of Delaware P, 2006, pp. 127–150.
Braziel, Jana Evans. *Caribbean Genesis: Jamaica Kincaid and the Writing of New Worlds.* State U of New York P, 2009.
Braziel, Jana Evans. "Daffodils, Rhizomes, Migrations: Narrative Coming of Age in the Diasporic Writings of Edwidge Danticat and Jamaica Kincaid." *Meridians: Feminism, Race, Transnationalism*, vol. 3, no. 2, 2003, pp. 110–131.
Bridges-Moussaron, Victoria. "Re-membering Diaspora: Punctuation, Prosopopoeia, Presence in Jamaica Kincaid's *Mr. Potter.*" *Le Tour Critique*, vol. 1, no. 1, 2013, n. pag., letourcritique.paris-nanterre.fr/index.php/letourcritique/article/view/8/html#bodyftn21. Accessed 18 Sep. 2018.
Brontë, Charlotte. *Jane Eyre.* 1847. Penguin, 2006.
Brophy, Sarah. "Angels in Antigua: The Diasporic of Melancholy in Jamaica Kincaid's *My Brother.*" *PMLA: Publications of the Modern Language Association of America*, vol. 117, no. 2, 2002, pp. 265–277.
Buckner, Brittnay. "Singular Beast: A Conversation with Jamaica Kincaid." *Callaloo*, vol. 31, no. 2, 2008, pp. 461–469.
Bush, Barbara. *Slave Women in Caribbean Society 1650–1838.* Heinemann Publishers, 1990.
Bush-Slimani, Barbara. "Hard Labour: Women, Childbirth and Resistance in British Caribbean Slave Societies." *History Workshop Journal*, vol. 36, 1993, pp. 83–99.
Butler, Judith. "The Body Politics of Julia Kristeva." *Hypatia: A Journal of Feminist Philosophy*, vol. 3, no. 3, Winter 1989, pp. 104–118.
Butler, Judith. *Gender Trouble: Feminism and the Subversion of Identity.* Routledge, 1990.
Butler, Judith. "Performative Acts and Gender Constitution." *Theatre Journal*, vol. 40, no. 4, 1988, pp. 519–531.
Castro, Ángela. "Power-in-Passivity: A Study of the Body and the Female Recognition in Jamaica Kincaid's *Lucy.*" *Journal of Commonwealth and Postcolonial Studies*, vol. 5, no. 2, 2017, pp. 19–28.
Caudeiron, Mable 'Cissie'. "From Livre to Douillette." *Our Island Culture*, edited by Lennox Honychurch, Dominica Cultural Council, 1982, pp. 31–36.
Chansky, Ricia Anne. "Between Selves: An Intertextual Approach to Jamaica Kincaid's *Among Flowers.*" *Biography: An Interdisciplinary Quarterly*, vol. 38, no. 1, 2015, pp. 135–151.
Chick, Nancy. "The Broken Clock: Time, Identity, and Autobiography in Jamaica Kincaid's *Lucy.*" *College Language Association Journal*, vol. 40, no. 1, 1996, pp. 90–103.
Classen, Constance. *Worlds of Sense: Exploring the Senses in History and Across Cultures.* Routledge, 1993.
Classen, Constance, et al. *Aroma: The Cultural History of Smell.* Routledge, 1994.
Cobham, Rhonda. "'Mwen na rien, Msieu': Jamaica Kincaid and the Problem of Caribbean Gnosis." *Callaloo*, vol. 25, no. 3, 2002, pp. 868–884.

Cook, Mariana. *Generations of Women: In Their Own Words.* Chronicle Books, 1998.
Coppola, Manuela. "Caribbean Autobiographies as Weapons in Identity Construction: Jamaica Kincaid's *Annie John.*" *The Caribbean Writer as Warrior of the Imaginary/ L'Ecrivain caribéen, guerrier de l'imaginaire*, edited by Kathleen Gyssels and Bénédicte Ledent, Rodopi, 2008, pp. 307–319.
Cousineau, Diane. "Autobiography: Innocent Pose or Naturalistic Lie? Jamaica Kincaid's Annie John." *Letters and Labyrinths: Women Writing/Cultural Codes*, U of Delaware P, 1997, pp. 118–127.
Cousineau, Diane. "Women and Autobiography: Is There Life Beyond the Looking Glass?" *Caliban: French Journal of English Studies*, vol. 31, 1994, pp. 97–105.
Covi, Giovanna. "Creolizing Cultures and Kinship: Then and There, Now and Here." *Perspectives of the Radical Other*, special issue of *Synthesis: An Anglophone Journal of Comparative Studies*, vol. 7, 2015, pp. 122–153.
Covi, Giovanna. "Jamaica Kincaid and the Resistance to Canons." *Out of the Kumbla: Caribbean Women and Literature*, edited by Carole Boyce Davies and Elaine Savory Fido, Africa World Press, 1990, pp. 345–354.
Cudjoe, Selwyn R. "Jamaica Kincaid and the Modernist Project: An Interview." *Callaloo*, vol. 39, 1989, pp. 396–411.
Dance, Daryl Cumber. *In Search of Annie Drew: Jamaica Kincaid's Mother and Muse.* U of Virginia P, 2016.
Dangarembga, Tsitsi. *Nervous Conditions.* 1998. Seal Press, 2004.
Daniels, Douglas Henry. *Pioneer Urbanites: A Social and Cultural History of Black San Francisco.* Temple UP, 1980.
Dash, J. Michael. *Édouard Glissant.* Cambridge UP, 1995.
Dash, J. Michael. Introduction. *Caribbean Discourse: Selected Essays*, by Édouard Glissant, translated by J. Michael Dash, U of Virginia P, 1999.
Davidson, Cathy N., editor. *The Oxford Companion to Women's Writing in the United States.* Oxford UP, 1995.
Davis, Angela. "Reflections on the Black Woman's Role in the Community of Slaves." *The Massachusetts Review*, vol. 13, no. 1–2, 1972, pp. 81–100.
De Ferrari, Guillermina. *Vulnerable States: Bodies of Memory in Contemporary Caribbean Fiction.* U of Virginia P, 2007.
De Souza, Pascale. "Creolizing Anancy: Signifyin(g) Processes in New World Spider Tales." *A Pepper-Pot of Cultures: Aspects of Creolization in the Caribbean*, edited by Gordon Collier and Ulrich Fleischmann, Rodopi, 2003, pp. 339–363.
Derrida, Jacques. "The Law of Genre." Translated by Avital Ronell, *Critical Inquiry*, vol. 7, no. 1, Autumn 1980, pp. 55–81.
Donnell, Alison. "When Writing the Other is Being True to the Self: Jamaica Kincaid's *The Autobiography of My Mother.*" *Women's Lives into Print: The Theory, Practice and Writing of Feminist Auto/Biography*, edited by Pauline Polkey, Palgrave, 1999, pp. 123–136.
Donnell, Alison, and Sarah Lawson Welsh, editors. *The Routledge Reader in Caribbean Literature.* Routledge, 1996.
Döring, Tobias. *Caribbean-English Passages: Intertextuality in a Postcolonial Tradition.* Routledge, 2002.
Douglas, Mary. *Purity and Danger: An Analysis of Concepts of Pollution and Taboo.* 1966. Routledge, 2002.
Dubois, Philippe. *L'Acte Photographique et Autres Essais.* Nathan, 1990.
Du Bois, W. E. B. *The Souls of Black Folk.* 1903. Dover, 1994.

Eakin, Paul John. *How Our Lives Become Stories: Making Selves.* Cornell UP, 1999.
Eco, Umberto. *Semiotics and the Philosophy of Language.* Indiana UP, 1984.
Edwards, Justin D. *Understanding Jamaica Kincaid.* U of South Carolina P, 2007.
Edwards, Justin D., and Rune Graulund. *Grotesque.* Routledge, 2013.
Eeva, Sara. "Whiteness and the Other Within: Creole Women in Jean Rhys's Caribbean Novels." *Close Encounters of an Other Kind: New Perspectives on Race, Ethnicity and American Studies*, edited by Roy Goldblatt et al., Joensuu University, 2005, pp. 247 – 259.
Emberley, Julia. "Material Fictions of Desire: Transactional Readings, Fashion, and the 'Worlding' of Everyday Life in Contemporary Women's Writing." *Fashion Theory*, vol. 11, no. 4, 2007, pp. 463 – 481.
Emery, Mary Lou. *Modernism, the Visual, and Caribbean Literature.* Cambridge UP, 2007.
Emery, Mary Lou. "Refiguring the Postcolonial Imagination: Tropes of Visuality in Writing of Rhys, Kincaid, and Cliff." *Tulsa Studies in Women's Literature*, vol. 16, no. 2, 1997, pp. 259 – 280.
Everett, Julin. "The Postcolonial Orphan's Autobiography: Authoring the Self in Jamaica Kincaid's *Mr. Potter* and Calixthe Beyala's *La Petite fille du réverbère.*" *College Literature*, vol. 36, no. 3, 2009, pp. 45 – 65.
Ewert, Jeanne C. "'Great Plant Appropriators' and Acquisitive Gardeners: Jamaica Kincaid's Ambivalent Garden (Book)." *Jamaica Kincaid and Caribbean Double Crossings*, edited by Linda Lang-Peralta, U of Delaware P, 2006, pp. 113 – 126.
Fabian, Johannes. *Time and the Other: How Anthropology Makes Its Object.* 1983. Columbia UP, 2002.
Fanon, Frantz. *Black Skin, White Masks.* Translated by Richard Philcox, Grove Press, 2008.
Feder, Helena. "Consuming Culture in *A Small Place* and *Among Flowers.*" *Ecocriticism and the Idea of Culture: Biology and the Bildungsroman*, Ashgate, 2014, pp. 94 – 130.
Ferguson, Moira. *Jamaica Kincaid: Where the Land Meets the Body.* UP of Virginia, 1994.
Ferguson, Moira. "A Lot of Memory: An Interview with Jamaica Kincaid." *The Kenyon Review*, vol. 16, no. 1, 1994, pp. 163 – 188.
Ferguson, Moira. "*Lucy* and the Landmark of the Colonizer." *Modern Fiction Studies*, vol. 39, no. 2, 1993, pp. 237 – 259.
Fernández Olmos, Margarite, and Lizabeth Paravisini-Gebert. *Creole Religions of the Caribbean: An Introduction from Vodou and Santería to Obeah and Espiritismo.* New York UP, 2003.
Figueredo, D. H., editor. *Encyclopedia of Caribbean Literature.* Greenwood Press, 2006.
Fjellestad, Danuta. "Towards an Aesthetics of Smell, or, the Foul and the Fragrant in Contemporary Literature." *Cauce: Revista de Filología y su Didáctica*, vol. 24, 2001, pp. 637 – 651.
Forbes, Curdella. "Fracturing Subjectivities: International Space and the Discourse of Individualism in Colin Channer's *Waiting in Vain* and Jamaica Kincaid's *Mr. Potter.*" *Small Axe: A Journal of Criticism*, vol. 25, 2008, pp. 16 – 35.
Forbes, Curdella. "Selling That Caribbean Woman down the River: Diasporic Travel Narratives and the Global Economy." *Journal of West Indian Literature*, vol. 13, no. 1 – 2, 2005, pp. 1 – 27.
Foucault, Michel. *The History of Sexuality. Vol I: An Introduction.* Translated by Robert Hurley, Vintage Books, 1990.
Foucault, Michel. "Lives of Infamous Men." *Power: The Essential Works of Michel Foucault 1954 – 1984*, translated by Robert Hurley et al., edited by James D. Faubion, New York Press, 2000, pp. 157 – 175.
Franklin, Ruth. *Shirley Jackson: A Rather Haunted Life.* Liveright Publishing, 2016.
Freud, Sigmund. *New Introductory Lectures on Psychoanalysis.* Translated by James Strachey, Norton, 1995.

Galassi, Jonathan. "Jamaica Kincaid & Jonathan Galassi: Authors and Editors in Conversation." *FSG: Work in Progress*, 7 Feb. 2013, fsgworkinprogress.com/2013/02/07/jamaica-kincaid-jonathan-galassi/. Accessed 18 May 2013.

Garber, Marjorie. *Vested Interests: Cross-Dressing and Cultural Anxiety.* 1992. Routledge, 2011.

Gardner-Corbett, Colena. "Escaping the Colonizer's Whip: The Binary Discipline." *College Language Association Journal*, vol. 49, no. 1, 2005, pp. 74 – 92.

Garis, Leslie. "Through West Indian Eyes." *The New York Times*, 7 Oct. 1990, www.nytimes.com/1990/10/07/magazine/through-west-indian-eyes.html?pagewanted=all&src=pm. Accessed 23 July 2013.

Gass, Joanne. "*The Autobiography of My Mother:* Jamaica Kincaid's Revision of *Jane Eyre* and *Wide Sargasso Sea.*" *Jamaica Kincaid and Caribbean Double Crossings*, edited by Linda Lang-Peralta, U of Delaware P, 2006, pp. 63 – 78.

Gates, Henry Louis, Jr. *The Signifying Monkey: A Theory of African-American Literary Criticism.* Oxford UP, 1988.

Gates, Henry Louis, Jr., and Nellie Y. MacKay, editors. *The Norton Anthology of African American Literature.* Norton, 1997.

Gauguin, Paul. "My Dear Vincent." Letter from Paul Gauguin to Vincent van Gogh. Le Pouldu, on or about Saturday, 28 June 1890, edited by Leo Jansen et al., 2009, Van Gogh Museum, b884 V/1962. vangoghletters.org/vg/letters/let892/letter.html. Accessed 21 Feb. 2017.

Gauguin, Paul. *Noa Noa.* Translated by O. F. Theis, Nicholas L. Brown, 1919.

Gauguin, Paul. *Noa Noa: Voyage De Tahiti.* 1901. Les Éditions G. Crès et Cie, 1924.

Geesaman, Lynn, and Jamaica Kincaid. *Poetics of Place: Photographs by Lynn Geesaman.* Umbrage Editions, 1998.

George, Verna Ena. "Putting up Resistance: Mother-Daughter Relationships in *Annie John* and *Nervous Conditions.*" *Journal of West Indian Literature*, vol. 7, no. 2, 1998, pp. 39 – 53.

Gersdorf, Catrin, and Juliane Braun, editors. *America After Nature: Democracy, Culture, Environment.* Universitätsverlag Winter, 2016.

Gervasio, Nicole. "The Power of the Weak Signifier: Wordsworth's Lucy in J. M. Coetzee's *Disgrace* and Jamaica Kincaid's *Lucy.*" *Modern Language Studies*, vol. 47, no. 2, 2018, pp. 36 – 57.

Gilbert, Sandra M., and Susan Gubar. *The Madwoman in the Attic: The Woman Writer and the Nineteenth-Century Literary Imagination.* Yale UP, 1984.

Gilmore, Leigh. *The Limits of Autobiography: Trauma and Testimony.* Cornell UP, 2001.

Gilmore, Leigh. "'What Was I?': Literary Witness and the Testimonial Archive." *Profession*, 2011, pp. 77 – 84.

Gilroy, Paul. *The Black Atlantic: Modernity and Double Consciousness.* Harvard UP, 1993.

Glissant, Édouard. *Caribbean Discourse: Selected Essays.* Translated by J. Michael Dash, U of Virginia P, 1999.

Glissant, Édouard. *Le discours antillais.* 1981. Gallimard, 1997.

Glissant, Édouard. *Poétique de la relation.* Gallimard, 1990.

Glissant, Édouard. *Poetics of Relation.* Translated by Betsy Wing, U of Michigan P, 2010.

Goff, Barbara E. Introduction. *Classics and Colonialism*, edited by Barbara E. Goff, Duckworth, 2005, pp. 1 – 24.

Graham, Mark. "Queer Smells: Fragrances of Late Capitalism or Scents of Subversion?" *The Smell Culture Reader*, edited by Jim Drobnick, Berg, 2006, pp. 305 – 319.

Graham, Maryemma, editor. *Cambridge Companion to the African American Novel.* Cambridge UP, 2004.

Gregg, Veronica Marie. "How Jamaica Kincaid Writes the Autobiography of Her Mother." *Callaloo*, vol. 25, no. 3, Summer 2002, pp. 920 – 937.
Grosz, Elizabeth. *Volatile Bodies: Towards a Corporeal Feminism.* Indiana UP, 1994.
Hardwick, Lorna. "Refiguring Classical Texts: Aspects of the Postcolonial Tradition." *Classics and Colonialism*, edited by Barbara E. Goff, Duckworth, 2005, pp. 107 – 117.
Harris, Wilson. "The Mystery of Timelessness." *The Caribbean Writer as Warrior of the Imaginary*, edited by Kathleen Gyssels and Bénédicte Ledent, Rodopi, 2008, pp. 25 – 30.
Hartman, Saidiya. "Venus in Two Acts." *Small Axe: A Journal of Criticism*, vol. 26, 2008, pp. 1 – 14.
Heermann, Ingrid. "Gauguins Tahiti: Anmerkungen aus ethnologischer Sicht." *Paul Gauguin. Tahiti*, edited by Christoph Becker, Verlag Gerd Hatje, 1998, pp. 147 – 165.
Heffernan, James. "Ekphrasis and Representation." *New Literary History*, vol. 22, 1991, pp. 297 – 316.
Hegel, Georg Wilhelm Friedrich. *Lectures on the Philosophy of History: Complete and Unabridged.* Newly translated by Ruben Alvarado, based on the 1857 translation by John Sibree. Wordbridge, 2011.
Hirsch, Marianne. *Family Frames: Photography, Narrative and Postmemory.* Harvard UP, 1997.
Hirsch, Marianne. *The Generation of Postmemory: Writing and Visual Culture After the Holocaust.* Columbia UP, 2012.
Hirsch, Marianne. "Resisting Images: Rereading Adolescence." *Provoking Agents: Gender and Agency in Theory and Practice*, edited by Judith K. Gardiner, U of Illinois P, 1995, pp. 249 – 279.
Holcomb, Gary. "Travels of a Transnational Slut: Sexual Migration in Kincaid's *Lucy*." *Critique: Studies in Contemporary Fiction*, vol. 44, no. 3, 2003, pp. 295 – 312.
Holcomb, Gary, and Kimberly. Holcomb. "Abusing the Colonial Desiring Machine: Sadomasochism in Jamaica Kincaid's *The Autobiography of My Mother*." *Journal of West Indian Literature*, vol. 8, no. 2, 1999, pp. 10 – 18.
Holland, Patricia. "'Sweet it is to scan…': Personal Photographs and Popular Photography." *Photography: A Critical Introduction*, edited by Liz Wells, 2nd ed., Routledge, 2000, pp. 117 – 164.
Hollmann, Eckhard. *Paul Gauguin: Bilder aus der Südsee.* Prestel, 1996.
hooks, bell. "In Our Glory: Photography and Black Life." *Picturing Us: African American Identity in Photography*, edited by Deborah Willis, New Press, 2000, pp. 44 – 53.
hooks, bell. "The Oppositional Gaze: Black Female Spectators." *Black Looks: Race and Representation*, South End Press, 1992, pp. 115 – 131.
hooks, bell. *Talking Back: Thinking Feminist, Thinking Black.* South End Press, 1989.
Hoving, Isabel. "Jamaica Kincaid Is Getting Angry." *In Praise of New Travelers: Reading Caribbean Migrant Women's Writing*, Stanford UP, 2001, pp. 184 – 237.
Hyde, Allan. "Offensive Bodies." *The Smell Culture Reader*, edited by Jim Drobnick, Berg, 2006, pp. 53 – 58.
"Hymn to Demeter." Homeric Hymns. Homeric Apocrypha. Lives of Homer. Edited and translated by Martin L. West. Harvard UP, 2003, pp. 33 – 71.
Jackson, Shirley. *Life Among the Savages.* 1953. Penguin Books, 1997.
Jackson, Tommie Lee. "The Symbolic Implications of the Portmanteau in Jamaica Kincaid's *Annie John*." *College Language Association Journal*, vol. 50, no. 3, 2007, pp. 298 – 311.
James, Louis. *Caribbean Literature in English.* Longman, 1999.
Jay, Julia De Foor. "'What a History You Have': Ancestral Memory, Cultural History, Migration Patterns, and the Quest for Autonomy in the Fiction of Jamaica Kincaid." *Middle Passages and the Healing Place of History: Migration and Identity in Black Women's Literature*, edited by Elizabeth Brown-Guillory, Ohio State UP, 2006, pp. 117 – 138.

Jefferson, Thomas. *Notes on the State of Virginia.* 1787. U of North Carolina P, 1954.
Jurney, Florence Ramond. "The Island and the Creation of (Hi)Story in the Writings of Michelle Cliff and Jamaica Kincaid." *Anthurium: A Caribbean Studies Journal*, vol. 4, no. 1, 2006, n. pag.
Kettler, Andrew. *The Smell of Slavery: Olfactory Racism and the Atlantic World.* Cambridge UP, 2020.
Kincaid, Jamaica. *Among Flowers: A Walk in the Himalaya.* National Geographic, 2005.
Kincaid, Jamaica. *Annie John.* 1985. Farrar, Straus and Giroux, 1997.
Kincaid, Jamaica. *At the Bottom of the River.* 1978. Farrar, Straus and Giroux, 2000.
Kincaid, Jamaica. *The Autobiography of My Mother.* Farrar, Straus and Giroux, 1996.
Kincaid, Jamaica, editor. *The Best American Essays 1995.* Houghton Mifflin, 1995.
Kincaid, Jamaica, editor. *The Best American Travel Writing 2005.* Houghton Mifflin, 2005.
Kincaid, Jamaica. "Biography of a Dress." *Grand Street*, vol. 43, 1992, pp. 92–100.
Kincaid, Jamaica. "Inside the American Snow Dome." *The Paris Review*, 17 Nov. 2020, www.theparisreview.org/blog/2020/11/11/inside-the-american-snow-dome/. Accessed 12 Dec. 2020.
Kincaid, Jamaica. *Lucy.* Farrar, Straus and Giroux, 1990.
Kincaid, Jamaica. *Mr. Potter: A Novel.* Farrar, Straus and Giroux, 2002.
Kincaid, Jamaica. *My Brother.* Farrar, Straus and Giroux, 1997.
Kincaid, Jamaica, editor. *My Favorite Plant: Writers and Gardeners on the Plants They Love.* Farrar, Straus and Giroux, 1998.
Kincaid, Jamaica. *My Garden (Book):.* Farrar, Straus and Giroux, 1999.
Kincaid, Jamaica. *Party: A Mystery.* Illustrated by Ricardo Cortés. Akashic Books, 2019.
Kincaid, Jamaica. *See Now Then.* Farrar, Straus and Giroux, 2013.
Kincaid, Jamaica. *A Small Place.* 1988. Farrar, Straus and Giroux, 2000.
Kincaid, Jamaica. *Talk Stories.* Farrar, Straus and Giroux, 2001.
Kincaid, Jamaica, and Eric Fischl. *Annie, Gwen, Lilly, Pam and Tulip.* Illustrated book with nine lithographs. Library Fellows of the Whitney Museum of American Art, 1986.
King, Rosamond. "Sex as Rebellion: A Close Reading of *Lucy* and *Brown Girl, Brownstones.*" *Journal of African American Studies*, vol. 12, no. 4, 2008, pp. 366–377.
Knepper, Wendy. "'How does your garden grow?' or Jamaica Kincaid's Spatial Praxis in *My Garden (Book):* and *Among Flowers: A Walk in the Himalaya.*" *Postcolonial Spaces: The Politics of Place in Contemporary Culture*, edited by Andrew Teverson and Sara Upstone, Palgrave Macmillan, 2011, pp. 40–56.
Koselleck, Reinhart. *Futures Past: On the Semantics of Historical Time.* Translated by Keith Tribe, Columbia UP, 2004.
Koselleck, Reinhart. *The Practice of Conceptual History: Timing History, Spacing Concepts.* Translated by Todd Samuel Presner et al., Stanford UP, 2002.
Koselleck, Reinhart. *Zeitschichten: Studien zur Historik.* Suhrkamp Verlag, 2000.
Kracauer, Siegfried. "Photography." Translated by Thomas Y. Levin, *Critical Inquiry*, vol. 19, no. 3, Spring 1993, pp. 421–436.
Krieger, Murray. *Ekphrasis: The Illusion of the Natural Sign.* Johns Hopkins UP, 1992.
Kristeva, Julia. *Powers of Horror: An Essay on Abjection.* Translated by Leon S. Roudiez, Columbia UP, 1982.
Kruger, Kathryn Sullivan. *Weaving the Word: The Metaphorics of Weaving and Female Textual Production.* Susquehanna UP, 2001.
Lacan, Jacques. "The Mirror Stage as Formative of the *I* Function as Revealed in Psychoanalytic Experience." *Écrits: The First Complete Edition in English.* Translated by Bruce Fink, W. W. Norton, 2006, pp. 75–81.

Lachmann, Renate. *Memory and Literature: Intertextuality in Russian Modernism.* Translated by Roy Sellars and Anthony Wall, U of Minnesota P, 1997.

Lang-Peralta, Linda, editor. *Jamaica Kincaid and Caribbean Double Crossings.* U of Delaware P, 2006.

Lang-Peralta, Linda. "'Smiling With My Mouth Turned Down': Ambivalence in Jamaica Kincaid's *Lucy* and *My Garden (Book):.*" *Jamaica Kincaid and Caribbean Double Crossings*, edited by Linda Lang-Peralta, U of Delaware P, 2006, pp. 33–44.

Largey, Gale, and Rod Watson. "The Sociology of Odors." *The Smell Culture Reader*, edited by Jim Drobnick, Berg, 2006, pp. 29–40.

Ledent, Bénédicte. "Voyages into Otherness: *Cambridge* and *Lucy.*" *Kunapipi: Journal of Postcolonial Writing*, vol. 14, no. 2, 1992, pp. 53–63.

Lejeune, Philippe. *On Autobiography.* Edited by Paul John Eakin, translated by Katherine Leary, U of Minnesota P, 1989.

Lima, Maria Helena. "Decolonizing Genre: Jamaica Kincaid and the Bildungsroman." *Genre: Forms of Discourse and Culture*, vol. 26, no. 4, 1993, pp. 431–459.

Lima, Maria Helena. "Imaginary Homelands in Jamaica Kincaid's Narratives of Development." *Callaloo*, vol. 25, no. 3, 2002, pp. 857–867.

Lyle, Timothy. "Prolonging 'Last Call': Jamaica Kincaid's Voyeuristic Pleasures in *My Brother.*" *Journal of West Indian Literature*, vol. 22, no. 1, 2013, pp. 33–62.

MacDonald-Smythe, Antonia. "Authorizing the Slut in Jamaica Kincaid's *At the Bottom of the River.*" *MaComère: Journal of the Association of Caribbean Women Writers and Scholars*, 1999, pp. 31–51.

Mahlis, Kristen. "Gender and Exile: Jamaica Kincaid's *Lucy.*" *Modern Fiction Studies*, vol. 44, no. 1, 1998, pp. 164–183.

Mårdberg, Maria. "'A Bleak, Black Wind': Motherlessness and Emotional Exile in Jamaica Kincaid's *The Autobiography of My Mother.*" *Reclaiming Home, Remembering Motherhood, Rewriting History: African American and Afro-Caribbean Women's Literature in the Twentieth Century*, edited by Verena Theile and Marie Drews, Cambridge Scholars Publishing, 2009, pp. 2–28.

Mardorossian, Carine. "Doubling, Healing, and Gender in Caribbean Literature." *MaComère: Journal of the Association of Caribbean Women Writers and Scholars*, vol. 12, no. 1, 2010, pp. 17–29.

Mardorossian, Carine. "Rape by Proxy in Contemporary Diasporic Women's Fiction." *Framing the Rape Victim: Gender and Agency Reconsidered*, Rutgers UP, 2014, pp. 112–128.

Marquis, Claudia. "'Not at Home in Her Own Skin': Jamaica Kincaid, History, and Selfhood." *Caliban: French Journal of English Studies*, vol. 21, 2007, pp. 101–114.

Mars-Jones, Adam. "A Very, Very Long Voyage Round My Father." Review of *Mr. Potter*, by Jamaica Kincaid. *The Guardian*, 28 July 2002, www.theguardian.com/books/2002/jul/28/fiction.features. Accessed 21 Aug. 2018.

Matos, Nicole C. "'The Difference between the Two Bundles': Body and Cloth in the Works of Jamaica Kincaid." *Callaloo*, vol. 25, no. 3, 2002, pp. 844–856.

Matos, Nicole C. "In the Vicinity of the Land of the Almost: The Stylistics of Jamaica Kincaid's *Mr. Potter.*" *The Journal of Commonwealth Literature*, vol. 44, no. 1, 2009, pp. 81–99.

Mbembe, Achille. *On the Postcolony.* U of California P, 2001.

McClintock, Anne. *Imperial Leather: Race, Gender and Sexuality in the Colonial Contest.* Routledge, 1995.

McClintock, Anne. "Maid to Order: Commercial Fetishism and Gender Power." *Social Text*, vol. 37, 1993, pp. 87–116.

McCormick, Stacie Selmon. "Navel-Erasing: Androgyny and Self-Making in Toni Morrison's *Song of Solomon* and Jamaica Kincaid's *The Autobiography of My Mother.*" *Imagining the Black Female*

Body: Reconciling Image in Print and Visual Culture, edited by Carol E. Henderson, Palgrave Macmillan, 2010, pp. 145 – 161.

McDonald-Smythe, Antonia. "*Kwik! Kwak!* Narrating the Self: A Reading of *The Autobiography of My Mother.*" *Journal of West Indian Literature*, vol. 8, no. 1, 1998, pp. 41 – 55.

Mirzoeff, Nicholas. *An Introduction to Visual Culture.* Routledge, 1999.

Mitchell, W. J. T. *Iconology: Image, Text, Ideology.* U of Chicago P, 1986.

Mitchell, W. J. T. "Imperial Landscape." *Landscape and Power*, edited by W. J. T. Mitchell, 2nd ed., U of Chicago P, 2002, pp. 5 – 34.

Mitchell, W. J. T. *Picture Theory: Essays on Verbal and Visual Representation.* U of Chicago P, 1994.

Moore-Gilbert, Bart. "A Concern Peculiar to Western Man? Postcolonial Reconsiderations of Autobiography as a Genre." *Postcolonial Poetics: Genre and Form*, edited by Patrick Crowley and Jane Hiddlestone, Liverpool UP, 2011, pp. 91 – 108.

Moore-Gilbert, Bart. "Western Autobiography and Colonial Discourse: An Overview." *Wasafiri: International Contemporary Writing*, vol. 21, no. 1, July 2006, pp. 9 – 16.

Morgan, Jennifer L. *Laboring Women: Reproduction and Gender in New World Slavery.* U of Pennsylvania P, 2004.

Morris, Kathryn E. "Jamaica Kincaid's Voracious Bodies. Engendering a Carib(bean) Woman." *Callaloo*, vol. 25, no. 3, Summer 2002, pp. 954 – 968.

Morrison, Toni. *Beloved.* 1987. Plume, 1988.

Murdoch, H. Adlai. "The Novels of Jamaica Kincaid: Figures of Exile, Narratives of Dreams." *Clockwatch Review: A Journal of the Arts*, vol. 9, no. 1 – 2, 1994, pp. 141 – 154.

Murdoch, H. Adlai. "Severing the (M)Other Connection: The Representation of Cultural Identity in Jamaica Kincaid's *Annie John.*" *Callaloo*, vol. 13, no. 2, 1990, pp. 325 – 340.

Murray, Melanie A. "Shifting Identities and Locations in Jamaica Kincaid's *My Garden (book):* and *A Small Place.*" *World Literature Written in English*, vol. 39, no. 1, 2001, pp. 116 – 126.

Nagel, James. "Desperate Hopes, Desperate Lives: Depression and Self-Realization in Jamaica Kincaid's *Annie John and Lucy.*" *Traditions, Voices, and Dreams: The American Novel Since the 1960s*, edited by Melvin J. Friedman and Ben Siegel, U of Delaware P, 1995, pp. 237 – 253.

Narain, Denise deCaires. "What Happened to Global Sisterhood? Writing and Reading 'the' Postcolonial Woman." *Third Wave Feminism: A Critical Exploration*, edited by Stacy Gillis et al., Palgrave Macmillan, 2004, pp. 240 – 251.

Narain, Denise deCaires. "Writing 'Home': mediating between 'the local' and 'the literary' in a selection of postcolonial women's texts." *Third World Quarterly*, vol. 26, no. 3, 2005, pp. 497 – 508.

Nasta, Susheila. "'Beyond the Frame': Writing a Life and Jamaica Kincaid's Family Album." *Contemporary Women's Writing*, vol. 3, no. 1, 2009, pp. 64 – 85.

"The National Dress." *Government of Saint Lucia Archival Website*, 1 Jan. 2013, archive.stlucia.gov.-lcsaint_lucia/the_national_dress.htm. Accessed 6 May 2015.

Naylor, Paul. *Poetic Investigations: Singing the Holes in History.* Northwestern UP, 1999.

Neumann, Birgit. "The Literary Representation of Memory." *Cultural Memory Studies: An International and Interdisciplinary Handbook*, edited by Astrid Erll et al., Walter de Gruyter, 2008, pp. 333 – 343.

Nichols, Jennifer. "'Poor Visitor': Mobility as/of Voice in Jamaica Kincaid's *Lucy.*" *MELUS: The Journal of the Society for the Study of the Multi-Ethnic Literature of the United States*, vol. 34, no. 4, Winter 2009, pp. 187 – 207.

Niesen de Abruna, Laura. "Family Connections: Mother and Mother Country in the Fiction of Jean Rhys and Jamaica Kincaid." *Motherlands: Black Women's Writing from Africa, the Caribbean, and South Asia*, edited by Susheila Nasta, Rutgers UP, 1991, pp. 257–289.
Nünning, Ansgar. "New Directions in the Study of Individual and Cultural Memory and Memory Cultures." *Fictions of Memory*, edited by Ansgar Nünning, special issue of *Journal for the Study of British Cultures*, vol. 10, no. 1, 2003, pp. 3–9.
Nyong'o, Tavia. "Unburdening Representation." *The Black Scholar*, vol. 44, no. 2, Summer 2014, pp. 70–80.
O'Brien, Susie. "The Garden and the World: Jamaica Kincaid and the Cultural Borders of Ecocriticism." *Mosaic*, vol. 35, no. 2, 2002, pp. 167–184.
O'Donnell, Rachel. "'This Is How to Make a Good Medicine to Throw Away a Child Before It Even Becomes a Child': The Maternal Voice in Jamaica Kincaid's 'Girl' and the Politics of Nature and Knowing in the Caribbean." *Journal of the Motherhood Initiative*, vol. 7, no. 1, 2016, pp. 100–110.
Omi, Michael, and Howard Winant. *Racial Formation in the United States.* 3rd ed., Routledge, 2014.
Orvell, Miles. *American Photography.* Oxford UP, 2003.
Paquet, Sandra Pouchet. *Caribbean Autobiography: Cultural Identity and Self-Representation.* U of Wisconsin P, 2002.
Paravisini-Gebert, Lizabeth. *Jamaica Kincaid: A Critical Companion.* Greenwood Press, 1999.
Paravisini-Gebert, Lizabeth. "Writers Playin' Mas': Carnival and the Grotesque in the Contemporary Caribbean Novel." *A History of Literature in the Caribbean. Vol. 3: Cross-Cultural Studies*, edited by Albert James Arnold, John Benjamins Publishing, 1997, pp. 215–236.
Parini, Jay, editor. *The Oxford Encyclopedia of American Literature.* Oxford UP, 2004.
Parks, Suzan-Lori. "From Elements of Style." *The America Play and Other Works*, Theatre Communications Group, 1995, pp. 6–18.
Patterson, Orlando. *Slavery and Social Death: A Comparative Study.* Harvard UP, 1982.
Pećić, Zoran. "Floral Diaspora in Jamaica Kincaid's Travel Writing." *Postcolonial Travel Writing: Critical Explorations*, edited by Justin D. Edwards and Rune Graulund, Palgrave Macmillan, 2011, pp. 138–155.
Peirce, Charles Sanders. "The Icon, Index, and Symbol." *Collected Papers Vol. II: Principles of Philosophy and Elements of Logic*, edited by Charles Hartshorne and Paul Weiss, Belknap P of Harvard U, 1960, pp. 156–173.
Pelt, April. "'Weary of Our Own Legacies': Rethinking Jane Eyre's Inheritance Through Jamaica Kincaid's *The Autobiography of My Mother.*" *ARIEL: A Review of International English Literature*, no. 3–4, 2010, pp. 73–90.
Perry, Donna. "Initiation in Jamaica Kincaid's *Annie John.*" *Contemporary American Women Writers: Gender, Class, Ethnicity*, edited by Lois Parkinson Zamora, Longman, 1998, pp. 128–137.
Perry, Donna. "An Interview with Jamaica Kincaid." *Reading Black, Reading Feminist: A Critical Anthology*, edited by Henry Louis Gates, Jr., Meridian-Penguin, 1990, pp. 492–509.
Petrie, Kathye Fetsko. "In Search of Shirley Jackson's House: Walking in Jackson's Footsteps in Bennington, Vermont." *Literary Hub*, 28 Sep. 2016, www.lithub.com/in-search-of-shirley-jacksons-house. Accessed 21 Aug. 2018.
Philip, M. NourbeSe. "The Absence of Writing or How I Almost Became a Spy." Afterword. *She Tries Her Tongue, Her Silence Softly Breaks*, 1989. Women's Press, 1993, pp. 76–91.
Pietrzak-Franger, Monika. "Silencing the Male: Rochester's Muteness." *NJES: Nordic Journal of English Studies*, vol. 7, no. 1, 2008, pp. 11–32.
Pindar. *Nemean Odes. Isthmian Odes. Fragments.* Translated by William H. Race, Harvard UP, 2006.

Rampaul, Giselle. "The West Indian Child as Subject/Object: Interrogating Notions of Power in *Annie John*." *Journal of Caribbean Literatures*, vol. 7, no. 1, 2011, pp. 153 – 160.

Réjouis, Rose-Myriam. "Caribbean Writers and Language: The Autobiographical Poetics of Jamaica Kincaid and Patrick Chamoiseau." *Massachusetts Review: A Quarterly of Literature, the Arts and Public Affairs*, vol. 44, no. 1 – 2, 2003, pp. 213 – 232.

Rhodes-Pitts, Sharifa. "Writing Between the Lines." Review of *Mr. Potter*, by Jamaica Kincaid. *The Women's Review of Books*, vol. 19, no. 12, 2002, p. 11.

Rhys, Jean. *Wide Sargasso Sea*. 1966. Penguin Books, 2001.

"Robert H. Woodworth, Biology Professor, 88." *The New York Times*, 6 Dec. 1990, www.nytimes.com/1990/12/06/obituaries/robert-h-woodworth-biology-professor-88.html. Accessed 7 Oct. 2014.

"Robert Woodworth." *Our Bennington*, Bennington College. 2014. our.bennington.edu/page/woodworth. Accessed 7 Oct. 2014.

Rodriguez, Maria Soledad. "A Little Miss in the Land of Little Women: Louisa May Alcott and Jamaica Kincaid." *Jamaica Kincaid and Caribbean Double Crossings*, edited by Linda Lang-Peralta, U of Delaware P, 2006, pp. 16 – 32.

Roe, Andrew. "The tics of Kincaid's prose fill the mind of 'Mr. Potter.'" Review of *Mr. Potter*, by Jamaica Kincaid. *SFGate*, 19 May 2002, www.sfgate.com/books/article/The-tics-of-Kincaid-s-prose-fill-the-mind-of-Mr-2820838.php. Accessed 16 Aug. 2018.

Roscher, Wilhelm Heinrich, editor. *Ausführliches Lexikon der griechischen und römischen Mythologie*. Vol. II.2. 1894 – 1897. Georg Olms Verlag, 1993.

Rostron, Nathan. "Jamaica Kincaid on writing and 'Outlaw American' culture." *USA Today*, 7 Mar. 2013, eu.usatoday.com/story/life/books/2013/03/07/jamaica-kincaid-interview/1971283/. Accessed 27 June 2013.

Ruchatz, Jens. "The Photograph as Externalization and Trace." *Cultural Memory Studies: An International and Interdisciplinary Handbook*, edited by Astrid Erll et al., Walter de Gruyter, 2008, pp. 367 – 378.

Rucker, Walter C. *The River Flows On: Black Resistance, Culture, and Identity Formation in Early America*. Louisiana State UP, 2006.

Russo, Mary. *Female Grotesque: Risk, Excess and Modernity*. Routledge, 1995.

Saal, Ilka. *Collusions of Fact and Fiction: Performing Slavery in the Works of Suzan-Lori Parks and Kara Walker*. U of Iowa P, 2021.

Schiebinger, Londa. *Plants and Empire: Colonial Bioprospecting in the Atlantic World*. Harvard UP, 2004.

Schultheis, Alexandra. "Family Matters in Jamaica Kincaid's *The Autobiography of My Mother*." *Jouvert: A Journal for Postcolonial Studies*, vol. 5, no. 2, Winter 2001, n. pag.

Sekula, Allan. "On the Invention of Photographic Meaning." *Thinking Photography*, edited by Victor Burgin, Macmillan, 1982, pp. 84 – 109.

Sela, Maya. "An Improbable Story, My Life." *Haaretz Daily Newspaper*, 18 June 2010, www.haaretz.com/2010-06-18/ty-article/an-improbable-story-my-life/0000017f-f5b6-d47e-a37f-fdbe5dbe0000. Accessed 13 Nov. 2012.

Sharrad, Paul. "Cloth and Self-Definition in Jamaica Kincaid's *The Autobiography of My Mother*." *Kunapipi: Journal of Postcolonial Writing*, vol. 26, no. 1, 2004, pp. 54 – 65.

Shiner, Larry. *Art Scents: Exploring the Aesthetics of Smell and the Olfactory Arts*. Oxford UP, 2020.

Sielke, Sabine. *Reading Rape: The Rhetoric of Sexual Violence in American Literature and Culture 1790 – 1990*. Princeton UP, 2002.

Simmel, Georg. *Soziologie: Untersuchungen über die Formen der Vergesellschaftung*. Duncker und Humblot, 1908.

Simmons, Diane. "Coming-of-Age in the Snare of History: Jamaica Kincaid's *The Autobiography of My Mother.*" *The Girl: Construction of the Girl in Contemporary Fiction by Women*, edited by Ruth O. Saxton, St. Martin's, 1998, pp. 107–118.

Simmons, Diane. *Jamaica Kincaid.* Twayne, 1994.

Simmons, Diane. "Jamaica Kincaid and the Canon: In Dialogue with Paradise Lost and Jane Eyre." *MELUS: The Journal of the Society for the Study of the Multi-Ethnic Literature of the United States*, vol. 23, no. 2, 1998, pp. 65–85.

Simmons, Diane. "The Rhythm of Reality in the Works of Jamaica Kincaid." *World Literature Today*, vol. 68, no. 3, 1994, pp. 466–472.

Smith, Mark. *How Race Is Made: Slavery, Segregation, and the Senses.* U of North Carolina P, 2006.

Smith, Sidonie, and Julia Watson. *Reading Autobiography: A Guide for Interpreting Life Narratives.* U of Minnesota P, 2001.

Smith, Sidonie, and Julia Watson. "Situating Subjectivity in Women's Autobiographical Practices." Introduction. *Women, Autobiography, Theory: A Reader*, edited by Sidonie Smith and Julia Watson, U of Wisconsin P, 1998, pp. 3–56.

Smith, William, editor. *A Dictionary of Greek and Roman Biography and Mythology.* Vol. II. 1880. Tauris, 2007.

Snodgrass, Mary Ellen. *Jamaica Kincaid: A Literary Companion.* McFarland and Company, 2008.

Sollors, Werner. *Neither Black nor White yet Both: Thematic Explorations of Interracial Literature.* Oxford UP, 1997.

Sontag, Susan. *On Photography.* 1977. Penguin, 2008.

Soto-Crespo, Ramón E. "Death and the Diaspora Writer: Hybridity and Mourning in the Work of Jamaica Kincaid." *Contemporary Literature*, vol. 43, no. 2, 2002, pp. 342–376.

Spillers, Hortense J. "Mama's Baby, Papa's Maybe: An American Grammar Book." *Diacritics: A Review of Contemporary Criticism*, vol. 17, no. 2, Summer 1987, pp. 64–81.

Spivak, Gayatri Chakravorty. "Thinking Cultural Questions in 'Pure' Literary Terms." *Without Guarantees: In Honour of Stuart Hall*, edited by Paul Gilroy et al., Verso, 2000, pp. 335–357.

Stanley, Liz. *The Auto/biographical I: The Theory and Practice of Feminist Auto/biography.* Manchester UP, 1992.

Stoler, Ann Laura. *Carnal Knowledge and Imperial Power: Race and the Intimate in Colonial Rule.* U of California P, 2002.

Stoler, Ann Laura. *Race and the Education of Desire: Foucault's History of Sexuality and the Colonial Order of Things.* Duke UP, 1995.

Strongman, Roberto. "Development and Same-Sex Desire in Caribbean Allegorical Autobiography: Shani Mootoo's *Cereus Blooms at Night*, and Jamaica Kincaid's *Annie John* and *Lucy.*" *Kunapipi: Journal of Postcolonial Writing*, vol. 27, no. 1, 2005, pp. 26–41.

Taylor, Diana. *The Archive and the Repertoire: Performing Cultural Memory in the Americas.* Duke UP, 2003.

Thieme, John. *Postcolonial Con-Texts: Writing Back to the Canon.* Continuum, 2001.

Thomson, Belinda. *Gauguin.* Thames and Hudson, 2001.

Tiffin, Helen. "The Metaphor of Anancy in Caribbean Literature." *Myth and Metaphor*, edited by Robert Sellick, Centre for Research in the New Literatures in English, 1982, pp. 15–52.

Tiffin, Helen. "'Replanted in This Arboreal Place:' Gardens and Flowers in Contemporary Caribbean Writing." *English Literatures in International Contexts*, edited by Heinz Antor and Klaus Stierstorfer, Universitätsverlag Winter, 2001, pp. 149–163.

Todorov, Tzvetan. “The Origin of Genres.” Translated by Richard M. Berrong, *New Literary History*, vol. 8, no. 1, Autumn 1976, pp. 159–170.

Turner, Patricia A. *Ceramic Uncles and Celluloid Mammies: Black Images and Their Influence on Culture.* Anchor Books, 1994.

Urry, John. *The Tourist Gaze.* 2nd ed., Sage, 2002.

Valens, Keja. “Obvious and Ordinary: Desire Between Girls in Jamaica Kincaid’s *Annie John.*” *Frontier: A Journal of Women Studies*, vol. 25, no. 2, 2004, pp. 123–149.

Vasquez, Shalene A. M. “In Her Own Image: Literary and Visual Representations of Girlhood in Toni Morrison’s *The Bluest Eye* and Jamaica Kincaid’s *Annie John.*” *The Griot: Journal of African American Studies*, vol. 31, no. 2, 2012, pp. 24–40.

Vorda, Allan, and Jamaica Kincaid. “An Interview with Jamaica Kincaid.” *The Mississippi Review*, vol. 20, no. 1–2, 1991, pp. 7–26.

Walcott, Derek. *Omeros.* Noonday Press, 1992.

Walker, Alice. “Looking for Zora.” *In Search of Our Mothers’ Gardens: Womanist Prose.* Harcourt, 1983, pp. 93–116.

Walters, Tracey L. *African American Literature and the Classicist Tradition: Black Women Writers from Wheatley to Morrison.* Palgrave Macmillan, 2007.

Ward, Frank Kingdon. *Plant Hunter’s Paradise.* Cape, 1937.

West, Elizabeth J. “In the Beginning There Was Death: Spiritual Desolation and the Search for Self in Jamaica Kincaid’s *Autobiography of My Mother.*” *South Central Review: The Journal of the South Central Modern Language Association*, vol. 20, no. 2–4, 2003, pp. 2–23.

West-Pavlov, Russell. *Temporalities.* Routledge, 2013.

Wexler, Laura. “Seeing Sentiment: Photography, Race, and the Innocent Eye.” *Female Subjects in Black and White: Race, Psychoanalysis, Feminism*, edited by Elizabeth Abel et al., U of California P, 1997, pp. 159–186.

White, Hayden. “Historiography and Historiophoty.” *The American Historical Review*, vol. 93, no. 5, 1988, pp. 1193–1199.

White, Hayden. *Metahistory: The Historical Imagination in Nineteenth-Century Europe.* Johns Hopkins UP, 1973.

Willis, Deborah. “A Search for the Self: The Photograph and Black Family Life.” *The Familial Gaze*, edited by Marianne Hirsch, UP of New England, 1999, pp. 107–123.

Wright, Michelle M. *Physics of Blackness: Beyond the Middle Passage Epistemology.* U of Minnesota P, 2015.

Young, Harvey. *Embodying Black Experience: Stillness, Critical Memory, and the Black Body.* U of Michigan P, 2010.

Index

 https://doi.org/10.1515/9783111027500-009

www.ingramcontent.com/pod-product-compliance
Ingram Content Group UK Ltd.
Pitfield, Milton Keynes, MK11 3LW, UK
UKHW041845031026
15587UKWH00003B/1018